Studies in Fuzziness and Soft Computing

Volume 409

The series "Studies in Fuzziness and Soft Computing" contains publications on various topics in the area of soft computing, which include fuzzy sets, rough sets, neural networks, evolutionary computation, probabilistic and evidential reasoning, multi-valued logic, and related fields. The publications within "Studies in Fuzziness and Soft Computing" are primarily monographs and edited volumes. They cover significant recent developments in the field, both of a foundational and applicable character. An important feature of the series is its short publication time and world-wide distribution. This permits a rapid and broad dissemination of research results.

Indexed by SCOPUS, DBLP, WTI Frankfurt eG, zbMATH, SCImago.

All books published in the series are submitted for consideration in Web of Science.

More information about this series at http://www.springer.com/series/2941

Peijia Ren · Zeshui Xu

Decision-Making Analyses with Thermodynamic Parameters and Hesitant Fuzzy Linguistic Preference Relations

Springer

Peijia Ren
School of Business Administration
South China University of Technology
Guangdong, China

Zeshui Xu
Business School
Sichuan University
Chengdu, China

ISSN 1434-9922 ISSN 1860-0808 (electronic)
Studies in Fuzziness and Soft Computing
ISBN 978-3-030-73252-3 ISBN 978-3-030-73253-0 (eBook)
https://doi.org/10.1007/978-3-030-73253-0

This Springer imprint is published by the registered company Springer Nature Switzerland AG
The registered company address is: Gewerbestrasse 11, 6330 Cham, Switzerland

Preface

Decision analysis has become an essential part of human activities. With the increasing complexity and uncertainty of practical problems, it is difficult for people to make accurate judgments for objects based on limited cognitions and incomplete information. In such a case, a fuzzy set that uses a membership function located in [0,1] to express the specific degree of an element belonging to a set was proposed by Zadeh (1965). It overcomes the weakness of traditional decision making in narrow results caused by only accurate judgments. It thus has been deeply studied in both qualitative and quantitative extensions, such as type-2 fuzzy set, intuitionistic fuzzy set, linguistic term set, and hesitant fuzzy linguistic term set. Besides proposing various sets to meet practical requirements, it is vital to introduce effective methods for handling decision making problems under uncertainty. Most decision making methods focus on adopting some transformation and fusion methods to get the results based on the numerical values of decision makers' judgments, which are limited in the full use of decision makers' opinions. For that reason, taking a complete account of both numerical values and judgment distributions can improve the effectiveness of the decision making procedure and get more scientific results. Based on this viewpoint, it is of theoretical significance and application value to introduce thermodynamic parameters (such as energy, exergy, and entropy) into decision making to construct methods for decision making problems, as initiated by Verma and Rajasankar (2017).

Furthermore, the lack of information makes decision makers depend on the pairwise comparisons among objects in some cases. Because the comparative mode accords with people's thinking and can directly reflect people's judgments, the methods for preference relations (PRs) become the leading way to solve decision making problems. To solve uncertain issues in reality, researchers have proposed different forms of PRs, such as intuitionistic PRs, hesitant PRs, linguistic PR, and hesitant fuzzy linguistic PR (HFLPR). By allowing decision makers to use continuous linguistic terms to describe their comparative results, the HFLPR gives a more applicable expression way and has good application prospects.

So far, it is worth discussing two questions based on the above descriptions:

(1) How to apply the thermodynamic parameters to make full use of decision makers' judgments in different decision scenarios;
(2) How to construct efficient procedures for decision making with HFLPRs (including the aspects of reasonable priority deriving model, valid consistency measure, and scientific group consensus process).

To this end, this book gives a thorough and systematic introduction to some research results on the above issues, which is organized into nine chapters:

Chapter 1 presents the background, aim, and significance of the book.

Chapter 2 lays the book's foundation by reviewing decision making methods based on thermodynamics and decision making methods based on HFLPRs, respectively.

Chapter 3 introduces a thermodynamic method for multi-criteria decision making (MCDM) in an intuitionistic fuzzy environment. It focuses on finding the thermodynamic statues of intuitionistic fuzzy numbers (IFNs) in the decision making process, defining the intuitionistic thermodynamic parameters, and applying the parameters' features to portray the numerical values and distributions of IFNs simultaneously. The chapter further demonstrates the comparison among the given method and the other two decision making methods through simulations, which manifests the impacts of the judgments' distributions on the final results. Moreover, a case concerning supporting hierarchical medical system is presented to illustrate the method.

Chapter 4 presents a thermodynamic method based on prospect theory for MCDM under a hesitant fuzzy environment. Firstly, a negative exponential function is utilized to transform the hesitant fuzzy decision matrix into the hesitant fuzzy prospect decision matrix to portray decision makers' psychological characteristics. Later on, the hesitant thermodynamic parameters are defined to describe the hesitant fuzzy prospect information from numerical values and distributions' perspectives. Accordingly, a decision making method is established by addressing the thermodynamic features of judgments and decision makers' psychological factors. The chapter further shows the simulations for discussing the validation and efficiency of the given method. Finally, it is illustrated by the case of decision making in the firing and exploding emergency event.

Chapter 5 demonstrates a thermodynamic method for heterogeneous MCDM problems with considering decision makers' confidence level of their judgments, which involves (1) extending the thermodynamic parameters into a heterogeneous environment, (2) revising the criteria weights that decision makers manifest more hesitancy to the judgments on alternatives, (3) introducing the TODIM method to collect the overall score for each option without transformation based on the heterogeneous thermodynamic information. Furthermore, this chapter performs simulations to discuss how the decision makers' sensitive attitudes towards uncertainty impact the decision making results. The given method is used to address a green supplier selection case under a low-carbon economy.

Chapter 6 gives a consistency definition for HFLPR, introduces a decision maker's satisfaction measurement to the consistency of an individual HFLPR, and constructs a mathematical programming model to derive priorities from a HFLPR. It highlights in (1) describing decision makers' uncertainty and fuzziness effectively, (2) portraying decision maker's satisfaction degree with the increasing marginal utility, and (3) applying the priority space to establish a mathematical programming model. Later on, simulations are provided to discuss the values of the model's undetermined parameters under different decision scenarios. Moreover, the chapter extends the model into an incomplete environment for improving the applicability. A case study concerning assessing the effects of hydropower stations' flood discharge and energy dissipation on the environment illustrates the given programming model.

Chapter 7 provides a group decision making method for HFLPRs based on individual modified extent measurement. The chapter firstly presents a kernel-based algorithm to cluster a group of HFLPRs and introduces a consensus procedure in each cluster based on measuring the modified extent of each HFLPR. Furthermore, the chapter investigates the impacts of decision makers' sensitive attitudes towards the distances between the individual HFLPRs and the overall HFLPR on the decision making results. Finally, the chapter illustrates the given method to select a flood discharge technique for a hydropower station.

Chapter 8 introduces a decision making method with consistent and consentaneous HFLPRs. Two key issues are being addressed: (1) defining a consistency index for HFLPRs and (2) establishing a consensus procedure to rationalize decision making results. This chapter further presents experiments to discuss consistency thresholds for different orders of HFLPRs instead of merely setting the consistency threshold as 0.9. Simulations are designed for manifesting the applicable conditions of the given method. The chapter further illustrates the given method to assess the erosion impacts of hydropower stations on the environment.

Chapter 9 concludes the book's primary work and demonstrates the future work based on the book's contents.

This book can be used as a reference for researchers and practitioners working in fuzzy mathematics, operations research, information science, management science, engineering, etc. It can also be used as a textbook for postgraduate and senior-year undergraduate students of the relevant professional institutions of higher learning.

This work was supported in part by the National Natural Science Foundation of China under Grant 72071135.

Special thanks to Prof. Xiao-Jun Zeng at the University of Manchester for lots of insightful suggestions.

Guangzhou, China
Chengdu, China
January 2021

Peijia Ren
Zeshui Xu

References

Zadeh, L. A.: Fuzzy sets. Inf. Control **8**, 338–356 (1965)
Verma, M., Rajasankar, J.: A thermodynamical approach towards group multi-criteria decision making (GMCDM) and its application to human resource selection. Appl. Soft Comput. **52**, 323–332 (2017)

Contents

Chapter 1
Introduction

1.1 Background

With the rapid development of the social economy and technology, today's management activities are not limited to the experience management mode but seek a more scientific and efficient management model. Usually, in management activities, we need to use effective and scientific means to evaluate each alternative according to the actual management situation, understand its corresponding advantages and disadvantages to make the best choice. The process, including defining the problem, determining the criteria values, and using appropriate methods to rank alternatives, is called decision making.

Decision analysis has always been an essential part of human activities. Decision makers often give judgments to alternatives according to their own experience in human beings' early decision making activities. With the continuous development of modern science and technology, decision science constructs a series of effective decision making methods depending on mathematics, computer science, economics, management science, etc., which improves the rationality of decision making results.

Different decision making problems and situations need to use different decision making methods to achieve scientificity. Therefore, adapting to the environment and establishing effective decision making mechanisms has taken people's urgent attention. With society's continuous progress in the past few decades, the complexity and uncertainty of decision making problems have been growing. In such a case, taking full account of decision making information and improving the accuracy of the information expression to build scientific decision making methods will enhance the decision making theory and make the results practical.

P. Ren and Z. Xu, *Decision-Making Analyses with Thermodynamic Parameters and Hesitant Fuzzy Linguistic Preference Relations*, Studies in Fuzziness and Soft Computing 409, https://doi.org/10.1007/978-3-030-73253-0_1

1.1.1 Development of Fuzzy Information

Considering that the decision making problem is complex, and people usually hard to make accurate judgments with their incomplete cognition and the preliminary information, a fuzzy set (FS) [33] was introduced to measure the fuzziness caused by various uncertain factors. It uses the membership function in the interval [0, 1] to express the specific degree of an element belonging to a set, which overcomes the weakness of traditional decision making in expressing accurate judgments. With society's diversification, the problems become complicated, and information is difficult to get nowadays. These cause the sharp rise of uncertainty in decision making, and FS sometimes is insufficient to be used to portray decision makers' minds. Much research has been addressed to extend the FS, which can be generally divided into two categories: quantitative extension and qualitative extension.

In terms of quantitative extension, Zadeh [34] proposed a type-2 fuzzy set to generally describe the uncertainty, which can measure the degree of a certain element belonging to a specific set and express the degree of uncertainty of the membership function. Later on, Atanassov [1] introduced the concept of intuitionistic fuzzy sets (IFS) to further cover more information under uncertainty by adding a non-membership function. Based on IFS, decision makers can evaluate objects from the perspectives of whether an element belongs to and does not belong to a set. Considering that decision makers may be hesitant in several possible fuzzy values when giving the membership degree, Torra [24] provided the concept of a hesitant fuzzy set (HFS) to retain all possible judgment values in the decision making process. Besides, based on the necessity of asymmetric description in some cases, Saaty's 1–9 scale [19] has been applied to re-express the above sets and come into being the sets of an intuitionistic multiplicative set [27], a hesitant multiplicative set [26], etc. There exist other quantitative extensions of FS such as fuzzy rough sets [4], fuzzy soft sets [12–14], Pythagorean fuzzy set [31, 32], neutrosophic number [21], etc.

In terms of qualitative extension, based on natural expressions, such as "high quality", "low quality", "good property", "poor performance", etc., Zadeh [34] proposed the concept of linguistic term set (LTS) to provide an efficient tool for depicting the decision making information in line with the actual expression habits of decision makers. As a set of linguistic representations based on the ordinal numbers, the LTS corresponds to people's daily language expression information to decision making problems and provides the basis for solving decision making problems with qualitative descriptions. After that, research on extending the LTS has been largely addressed. To apply the linguistic expression into the knowledge system, Torra [23] proposed a semantic negation function of the ordered linguistic scale. Herrera and Herrera-Viedma [7, 8] offered a decision analysis for linguistic information based on operator and aggregation. These works have made in-depth research on the asymmetric and symmetric distributions of LTSs. To prevent information loss and ensure the accuracy of decision making results, Herrera and Martínez [9] introduced the concept of 2-tuple fuzzy linguistic information and studied its aggregation techniques. Based on this work, Dong et al. [3] extended the expression of 2-tuple fuzzy linguistic information into the numerical scale and constructed the optimization model for the numerical scale of the linguistic term set. Besides, Xu [28] proposed a virtual linguistic term to transform the discrete linguistic information into continuous linguistic information for increasing the applicability. The sets mentioned above manifest limitations in some cases. For example, due to the complexity existing in the decision making problems, it is difficult for decision makers to express their opinions by a single linguistic term. The linguistic terms such as "above average", "between good and very good", etc., are commonly used in practice. Based on this situation, Rodríguez et al. [18] proposed a hesitant fuzzy linguistic term set (HFLTS). It allows decision makers to utilize several continuous linguistic terms for evaluating, which increases the flexibility of the qualitative description. Pang et al. [17] further studied the extension of HFLTS and gave the concept of probabilistic linguistic term set (PLTS) with containing the importance of each possible linguistic term in a set.

1.1.2 Development of Hesitant Fuzzy Linguistic Preference Relation

The lack of available information makes decision makers depend on the pairwise comparison between objectives in some decision making environments. As the pairwise comparison accords with the way of thinking and can directly reflect decision makers' judgments, decision making with preference relations (composed by pairwise comparisons) has become the main branch of modern decision making. The concept of preference relation was initialed in [19, 20] to describe the preference degree of an objective over another.

The uncertainty and complexity of the decision making environment inevitably lead to preference relation with a fuzzy mind. Based on the concept of FS [33], a fuzzy preference relation (FPR) was proposed for portraying decision makers' preferences with fuzziness [16, 22]. Later on, seeking a more comprehensive judgment way and achieving more effective decision making results, [30] introduced intuitionistic fuzzy preference relation (IFPR), which simultaneously depicts the preference degree from the perspectives that an objective is superior to and inferior to another. Considering that it is difficult for decision makers to describe their preferences with only one membership degree in some cases, Xia and Xu proposed a hesitant fuzzy preference relationship (HFPR) [26] to record all possible preferences on objectives. Moreover, preference relations based on other types of fuzzy were provided for different decision making scenarios, including triangular fuzzy preference relation [25], intuitionistic multiplicative preference relation [27], hesitant multiplicative preference relation [35], etc.

Considering that decision makers are accustomed to using linguistic terms (LTs) to demonstrate their opinions under uncertainty, the linguistic preference relation (LPR) was proposed to measure the preference degree between two objectives through "good", "fair", "poor" and other terms [5–8, 29]. With the increasing complexities in decision making, the increasing information cost in management activities, and resource limitation, the LPR shows defects in capturing decision makers' judgments. Then, based on the HFLTS [18], Zhu and Xu [36] gave the concept of hesitant fuzzy linguistic preference relation (HFLPR), which allows decision makers to use several consecutive linguistic terms to describe the results of pairwise comparisons. It provides a form of information that conforms to decision makers' description habit to express their judgments on objectives when facing uncertainty and complexity.

1.1.3 Development of Thermodynamics

Thermodynamics is a subject that explores the heat in substance and the interaction between substances. The thermodynamics was initiated to measure a substance's temperature by Jean Re, Robert Hooke, Anders Celsius, and other researchers. Later on, through experiments, Johann Heinrich Lambert, Isaac Newton, and Joseph black pointed out that temperature and substance are different concepts. On this basis, Irvine put forward the idea of specific heat capacity, which establishes thermodynamics [2, 15].

Classical thermodynamics is related to the study of the principles of heat. The relevant research led to three thermodynamics laws, namely, The First Law Of Thermodynamics proposed by James Prescott Joule, The Second Law Of Thermodynamics proposed by Rudolph Clausius and Lord Kelvin, and The Third Law Of Thermodynamics proposed by Nernst. These three laws explain the physical phenomenon, such as the form of energy, the transformation of energy, and the conditions of energy transformation when molecules move [11]. Furthermore, the research on the three laws of thermodynamics has been further developed. For example, the concept of entropy [10] was proposed to be an irreversible material state parameter to describe the part of energy wasted and unusable in the process of spontaneous energy transformation.

After that, by combining thermodynamics with molecular dynamics and quantum mechanics, the research topics in modern physics, such as statistical thermodynamics, quantum statistical mechanics, etc., were put forward [11].

1.2 Importance of Decision-Making Analyses with Thermodynamic Parameters and Hesitant Fuzzy Linguistic Preference Relations

With the diversification of social life, the randomness and uncertainty of decision making problems increase gradually. Therefore, the uncertain decision making framework also needs to develop with the evolution of society. Nowadays, the development of the world economy and information technology have brought a leap forward to our social life, but meanwhile, they have brought challenges to our decision making activities. The decision making environment, mode, object, scale, and other aspects are changing with society's information, i.e., the problems we are facing now include many unknown and complex factors. Under the circumstances, it is significant to improve decision making accuracy by making full use of the obtained decision making information and introducing practical information expression in decision making.

The existing decision making methods mainly rely on the numerical value of decision making information on addressing the numerical transformation and fusion, and then get the ranking of objectives. Due to the increasing randomness and complexity of decision making, it is difficult to obtain useful decision making information. Therefore, how to improve the existing decision making methods and fully make use of decision making information is of great significance to ensure the effectiveness of decision making results. Thermodynamics, as a discipline describing the substance energy and the storage or transformation form in a system, can reflect its quantity and internal structure. Therefore, introducing thermodynamics into the decision making field can represent the numerical value and distribution characteristics of decision making information and is a pivotal issue to be discussed.

Since HFLPR can provide information expression for decision makers based on their description habits, decision making with HFLPRs has become an excellent way to solve uncertain decision making problems. At present, it is still a challenge to achieve a well-established theory for decision making with HFLPRs, limiting the application of HFLPRs. To improve the decision theory of HFLPRs, the issues such as building models with the consistency of HFLPRs, collecting group wisdom for obtaining scientific solutions based on HFLPRs, considering different cognitions of decision makers in the group consensus model, etc., are necessary to be discussed.

Based on the above analyses, addressing the problems presented helps enhance decision making accuracy in various ways. It can provide new ideas for decision making methods under uncertain environments, which is the inevitable demands of uncertain decision making.

1.3 Aim of Book

The aim of the book can be briefly introduced as considering the uncertain and complex factors existing in the decision making problems, as well as the limitations of decision makers' knowledge and cognition, the book introduces decision making methods from the perspectives of making full use of decision making information and applying practical information expressions. Specifically,

(1) Under the decision making environment of intuitionistic fuzzy, hesitant fuzzy, and heterogeneous information, introduce the methods based on bringing in thermodynamic parameters into the decision making process with considering the bounded rationality and limited cognitions of decision makers.

(2) Construct a mathematical programming model based on the consistent property of HFLPRs to drive the priority vector from it.
(3) Consider the individual cognitions of decision makers, establish consensus models for group decision making with HFLPRs from the perspectives of the individual adjustment measurement of decision makers and the consistency of individual HFLPRs, respectively.
(4) Simulate applying the above methods in hierarchical diagnosis and treatment decisions, emergency plan selection, flood discharge and energy dissipation selection, etc.

References

1. Atanassov, K.T.: Intuitionistic fuzzy set. Fuzzy Sets Syst. **20**, 87–96 (1986)
2. Bozsaky, D.: The historical development of thermodynamics. Acta Technica Jaurinensis, 3(3–15), (2010)
3. Dong, Y.C., Xu, Y.F., Yu, S.: Computing the numerical scale of the linguistic term set for the 2-tuple fuzzy linguistic representation model. IEEE Trans. Fuzzy Syst. **17**(6), 1366–1378 (2009)
4. Dubois, D., Prade, H.: Rough fuzzy sets and fuzzy rough sets. Int. J. Gen. Syst. **17**(2–3), 191–209 (1990)
5. Fan, Z.P., Chen, X.: Consensus Measures and Adjusting Inconsistency of Linguistic Preference Relations in Group Decision Making. Springer-Verlag, Germany (2005)
6. Herrera-Viedma, E., Herrera, F., Chiclana, F.: A consensus model for multiperson decision making with different preference structures. IEEE Trans. Syst. Man Cybernet. Part A: Syst. Humans **32**(3), 394–402 (2002)
7. Herrera, F., Herrera-Viedma, E.: Choice functions and mechanisms for linguistic preference relations. Eur. J. Oper. Res. **120**, 144–161 (2000)
8. Herrera, F., Herrera-Viedma, E.: Linguistic decision analysis: steps for solving decision problems under linguistic information. Fuzzy Sets Syst. **115**(1), 67–82 (2000)
9. Herrera, F., Martínez, L.: A 2-tuple fuzzy linguistic representation model for computing with words. IEEE Trans. Fuzzy Syst. **8**(6), 746–752 (2000)
10. https://en.wikipedia.org/wiki/Entropy
11. https://en.wikipedia.org/wiki/Thermodynamics
12. Maji, P.K., Roy, A.R., Biswas, R.: Fuzzy soft sets. J. Fuzzy Math. **9**(3), 589–602 (2001)
13. Maji, P.K., Roy, A.R., Biswas, R.: An application of soft sets in a decision making problem. Comput. Math. Appl. **44**(8–9), 1077–1083 (2002)
14. Maji, P.K., Roy, A.R., Biswas, R.: Soft set theory. Comput. Math. Appl. **45**(4–5), 555–562 (2003)
15. Müller, I.: A History of Thermodynamics-The Doctrine of Energy and Entropy. Springer, Berlin (2007)
16. Orlorski, S.A.: Decision making with a fuzzy preference relation. Fuzzy Sets Syst. **1**(3), 155–167 (1978)
17. Pang, Q., Wang, H., Xu, Z.S.: Probabilistic linguistic term sets in multi-attribute group decision making. Inf. Sci. **369**, 128–143 (2016)
18. Rodríguez, R.M., Martínez, L., Herrera, F.: Hesitant fuzzy linguistic terms sets for decision making. IEEE Trans. Fuzzy Syst. **20**(1), 109–119 (2012)
19. Saaty, T.L.: A scaling method for priorities in hierarchical structures. J. Math. Psychol. **15**(3), 234–281 (1977)
20. Saaty, T.L.: The Analytic Hierarchy Process. McGraw-Hill, New York (1980)

21. Smarandache, F.: Introduction to Neutrosophic Measure Neutrosophic Integral and Neutrosophic Probability. Romania, Sitech and Education Publisher, Craiova (2013)
22. Tanino, T.: Fuzzy preference orderings in group decision making. Fuzzy Sets Syst. **12**(2), 117–131 (1984)
23. Torra, V.: Negation functions based semantics for ordered linguistic labels. Int. J. Intell. Syst. **11**, 975–988 (1996)
24. Torra, V.: Hesitant fuzzy sets. Int. J. Intell. Syst. **25**(6), 529–539 (2010)
25. Van Laarhoven, P.J.M., Pedrycz, W.: A fuzzy extension of Saaty's priority theory. Fuzzy Set Syst. **11**, 229–241 (1983)
26. Xia, M.M., Xu, Z.S.: Managing hesitant information in GDM problems under fuzzy and multiplicative preference relations. Int. J. Uncertain. Fuzziness Knowl.-Based Syst. **21**(06), 865–897 (2013)
27. Xia, M.M., Xu, Z.S., Liao, H.C.: Preference relations based on intuitionistic multiplicative information. IEEE Trans. Fuzzy Syst. **21**(1), 113–133 (2012)
28. Xu, Z.S.: A method based on linguistic aggregation operators for group decision making with linguistic preference relations. Inf. Sci. **166**(1–4), 19–30 (2004)
29. Xu, Z.S.: Deviation measures of linguistic preference relations in group decision making. Omega **33**(3), 249–254 (2005)
30. Xu, Z.S.: Intuitionistic preference relations and their application in group decision making. Inf. Sci. **177**, 2363–2379 (2007)
31. Yager, R.R.: Pythagorean fuzzy subsets. In: Proceedings of The Joint IFSA Wprld Congress and NAFIPS Annual Meeting. Edmonton, Canada (2013)
32. Yager, R.R.: Pythagorean membership grades in multicriteria decision making. IEEE Trans. Fuzzy Syst. **22**, 958–965 (2014)
33. Zadeh, L.A.: Fuzzy sets. Inf. Control **8**, 338–356 (1965)
34. Zadeh, L.A.: The concept of a linguistic variable and its application to approximate reasoning-part I. Inf. Sci. **8**(3), 199–249 (1975)
35. Zhu, B., Xu, Z.S.: Analytic hierarchy process-hesitant group decision making. Eur. J. Oper. Res. **239**(3), 794–801 (2014)
36. Zhu, B., Xu, Z.S.: Consistency measures for hesitant fuzzy linguistic preference relations. IEEE Trans. Fuzzy Syst. **22**, 34–45 (2014)

Chapter 2
Literature Review

2.1 Review of Decision-Making Based on Thermodynamics

Most of the existing decision making methods with thermodynamics focus on the application of entropy. Given this, here, we review the decision making theory based on the entropy concept.

2.1.1 Decision Methods with Criteria Entropy

Entropy is a tool to describe the effectiveness of a decision making system, which is usually processed in building methods to obtain criteria weights in decision making. Some achievements have been made in the relevant research. To avoid the uncertainty of decision makers' subjective judgments, Chen and Wang [4] constructed an entropy weight method to determine the weights of criteria with qualitative and quantitative information in decision making, and combined the TOPSIS (technical for order of preference by similarity to internal solution) to rank alternatives. Based on the ordered weighted average operator, Yager [116] established a model for deriving the priorities based on maximizing the entropy in the decision making process.

In addition, the research on decision analysis with grey system theory and entropy weight method has been discussed. Luo and Liu [59] gave a decision making method based on grey fuzzy relation based on optimization theory and the entropy maximization principle. They further provided a decision making algorithm for the situation of known criteria weights and unknown criteria weights, respectively. Mi et al. [68] studied a way to obtain the weight information for criteria based on the entropy definition and proposed a decision making method with grey fixed weight clustering based on the entropy weight. For handling the decision making problem with incomplete information, Luo and Li [60] presented a way based on entropy weight approach and grey correlation analysis, which can not only overcome the incomplete criteria

P. Ren and Z. Xu, *Decision-Making Analyses with Thermodynamic Parameters and Hesitant Fuzzy Linguistic Preference Relations*, Studies in Fuzziness and Soft Computing 409, https://doi.org/10.1007/978-3-030-73253-0_2

weights caused by limited thinking of decision makers, but also makes full use of decision making information and improve the effectiveness of decision making.

The complexity and uncertainty existing in decision making problems lead to different kinds of fuzzy information expression during the decision making process, then the entropy weight determination under different fuzzy environments has been widely studied. Based on the entropy proposed by Shannon as the primary way to obtain criteria weights, Lotfi and Fallahnejad [58] provided a method for determining entropy weight for interval information and fuzzy information. After that, Xu and Xia [113] introduced the concepts of entropy and cross entropy into a hesitant fuzzy environment and proposed MCDM methods based on hesitant fuzzy entropy weight and hesitant fuzzy cross entropy weight, respectively. For the decision making problem with triangular fuzzy information, Garg et al. [16] put forward to derive the criteria weights by entropy function. Tian et al. [85] proposed to utilize fuzzy cross entropy to calculate the deviation between an alternative and an ideal one for handling interval neutral information and then constructed two optimization models to determine the criteria weights. He et al. [25] solved the decision making problem with linguistic information by establishing a model based on the linguistic aggregation operator and entropy weight-deriving method. Gou et al. [22] gave the entropy measure and cross entropy measure under a hesitant fuzzy linguistic environment, and then presented a model to determine the criteria weights.

Besides, the decision making with entropy weight under an intuitionistic fuzzy environment has been largely discussed. Firstly, Xu and Hu [111] proposed the entropy-based weight-deriving process by constructing the score matrix of intuitionistic fuzzy information. Based on defining the intuitionistic fuzzy entropy weight, Wu and Zhang [92] introduced several entropy weight measurements, such as Szmidt and Kacprzyk entropy, weighted De Luca-Terminal entropy, the weighted entropy based on score function and minimum maximization, etc. They further established an optimization model for generating the criteria weights according to the principle of minimum entropy. After that, the concepts of entropy and cross entropy were developed into an intuitionistic fuzzy environment [98]. For the intuitionistic fuzzy decision making with unknown or partial known criteria weights, Liu and Ren [55] studied an entropy measurement, then built an optimization model based on the minimum entropy principle and the extended entropy weight method. Ye [120] provided an entropy measurement method for interval-valued intuitionistic fuzzy sets (IVIFS), and gave an entropy weight model to obtain criteria weights. Later on, by analogy with intuitionistic fuzzy cross entropy, Ye [121] presented the cross entropy measurement of IVIFSs based on the ideal and non-ideal measures, and constructed objective programming for obtaining the weights of alternatives. To evaluate the criteria weights in the decision making problems with IVIFSs, researchers have established different models based on various entropy measures, such as the information entropy, the entropy for decision matrix credibility, and the continuous weighted entropy [35, 74, 123]. In view of group decision making with unknown criteria and expert weights under an interval-valued intuitionistic fuzzy environment, Meng and Tang [66] established a mathematical model to obtain the optimal fuzzy metric by using cross entropy measure and Shapley function. Then, Qi et al. [75]

constructed a maximization optimization model for getting unknown criteria weights on the basis of generalized cross entropy measure of IVIFSs, meanwhile, derived the importance degrees of decision makers based on creating a comprehensive algorithm.

Furthermore, the application of entropy in decision making problems with preference relations has been studied. For improving the effectiveness of the traditional analytic hierarchy process (AHP) in a fuzzy environment, Mon et al. [70] proposed a triangular fuzzy AHP based on the definition of entropy weight and applied it to evaluate the weapon system. To enhance the performance of the decision making system, Chen and Qu [7] developed a fuzzy decision making model based on entropy technology to modify the weight information in the fuzzy AHP by using the Delphi method. Based on the relative entropy, Chen and Zhou [3] established a programming model to minimize the difference between the individual priority and the group priority to handle the group decision making with interval reciprocal preference relations.

2.1.2 Decision-Making with Entropy Applications

The applications of entropy in the decision tree, probability decision, and system instability have widely been studied, which can be briefly introduced as:

(1) For the entropy-based decision tree model, Ichihashi et al. [32] utilized the uncertainty measure of maximizing entropy into interactive Dichotomizer3 under a fuzzy environment to avoid the loss of decision makers' opinions due to simplifying the decision tree. With computer assistance, Lee and Yang [42] proposed a decision tree construction model based on entropy feature extraction to diagnose breast cancer. To overcome the defect that the existing monotone algorithms are sensitive to the number of samples and improve the robustness of the decision tree, Hu et al. [31] applied Shannon's entropy to the extraction of ordinal structured data, and successively constructed a decision tree algorithm based on rank mutual information.

(2) The probability decision model's main idea is to introduce the entropy to handle the probabilities of criteria and alternatives in uncertain decision making problems. Smith [78] proposed using the entropy function to measure the uncertainty of the probability distribution based on evaluating the occurrence probability of the alternatives. In addition, Buckley [2] established decision making models for the two cases: (a) only the prior distribution of alternatives' occurrence probabilities is known; (b) the prior distribution of alternatives' occurrence probabilities is known; meanwhile, the constraints of alternatives' occurrence probabilities are known.

(3) The uncertainty measurement of information has become an important aspect in decision analysis since it can give a new angle for analyzing data. To this end, some achievements in measuring the uncertainty of the decision system through entropy have been made. Considering that the existing heuristic algorithms

based on the rough set theory are time-consuming in dealing with large-scale incomplete decision making systems, Sun et al. [80] introduced entropy to measure the uncertainty of rough sets. Dai et al. [9] provided the probability of rough decision entropy to evaluate the uncertainty of the interval-valued decision system.

Furthermore, Myung [72] used entropy to interpret the decision constraint model and the decision text model. He further pointed out that several forms of decision constraints in the decision constraint model can be made as the maximum entropy solution of the model for the information with relatively limited structure. Szmidt and Kacprzyk [77] developed a non-probabilistic-type entropy to demonstrate the geometric description of the IFS. To deal with the decision making problem with risk, Yang and Qiu [117] gave the utility entropy risk measurement to reflect the individual's intuitive attitude to risk and built a decision making model based on expected utility and entropy. It extends the classical decision making model under risk to a more general situation. A method for obtaining the maximum entropy principle of joint probability distribution was analyzed based on low-order joint probability assessment in an uncertain environment [1]. Ye [119] presented the difference between the optimal alternative and the ideal alternative with IFSs by the concept of cross entropy. Along the same line, Zhao et al. [131] redefined the relative entropy between an alternative and the ideal alternative, the relative entropy between an alternative and the negative-ideal alternative, then proposed a ranking method for alternatives. Montesarchio et al. [71] proposed a utility function minimization model based on the information entropy for implementing a flood monitoring system. Then, they analyzed the system performance of the model in the aspects of a proper release, wrong release, and no alarm. Hsu [30] combined the financial ratio variables with the risk-adjusted capital return and constructed a factor analysis and entropy-based TOPSIS model for investment decision making. Dai et al. [8] applied Shannon's entropy to criteria selection in incomplete decision making systems. After that, Ye [122] gave the cross entropy of a single-valued neutrosophic set as an extension of FS's cross entropy and proposed a decision making method with the proposed cross-entropy. With the idea that a heuristic feature selection algorithm based on rough set is a useful feature selection tool, Jiang et al. [33] extended Shannon's information entropy of rough set and established a relative decision entropy-based model.

Recently, Verma and Rajasankar [86] introduced the thermodynamic parameters, such as energy, exergy, and entropy, into MCDM problems to describe the availability and unavailability of decision information. They moved forward to constructing a group decision making method based on thermodynamic parameters with numerical values and triangular fuzzy numbers. The method can fully consider the numerical value and distribution of all decision makers' decision information in the decision making process and ensure the reliability of decision making results.

2.2 Review of Decision-Making Based on Hesitant Fuzzy Linguistic Preference Relation

2.2.1 Decision Models for Deriving Priority Vector

The decision making problem based on preference relation seeks to obtain the priority vector of alternatives based on decision makers' preference relations.

The preference information given by decision makers is often fuzzy with the limitation of information and knowledge. So far, the theory of deriving priority vector from fuzzy preference relations (FPR) has been studied. Xu and Da [110] proposed to use the minimum deviation to obtain the priority vector from a FPR. To get the priority vector for a FPR and a multiplicative preference relation (MPR), Fan et al. [14] established a goal programming model based on minimizing the difference between the preference information and the corresponding priority information. To address the same issue, Wang et al. [90] constructed a mathematical programming model by using chi-square distribution on the basis of describing the difference between a FPR/MPR and the corresponding priority information. Later on, the least-square method was applied to acquire the priority vector from a FPR with incomplete information [18]. Based on the definition of additive consistent of a FPR, Wang et al. [88] provided linear programming models for generating priorities from additively consistent or inconsistent interval FPR. By addressing the consistency and the interval priorities simultaneously, Meng et al. [65] gave two consistency-based linear programming models to derive priorities from an interval FPR.

Besides, there exist various priority derivation models for more general preference relations, some of which can be summarized as follows:

(1) For the IFPR, Xu [107] first proposed to use intuitionistic fuzzy arithmetic average operator and intuitionistic fuzzy weighted average operator to obtain the priority of each alternative based on an IFPR. By converting an IFPR into an interval-valued preference relation, Xu [108] studied a method based on error analysis to derive the priority vector of alternatives. Considering that the consistency property is very significant for ensuring the logic of decision making, Xu [106] connected the relationship between the original IFPR and its corresponding score matrix, studied the additive consistency and multiplicative consistency of an IFPR, and established a series of linear programming models based on the inconsistency minimization measure. Later on, Meng et al. [62] created a 0–1 mixed programming model to obtain the priority from an IFPR by measuring the multiplicative consistency of the mixed matrix based on the IFPR and its dual form. Based on the assumptions of consistency and preference variables obeying the uncertainty distribution, Gong et al. [20, 21] built an optimal weighting model based on linear uncertain constraints for IFPR. Gong et al. [20, 21] utilized non-linear utility functions into the optimization ordering model for IFPRs, which results in a priority vector of utility or deviation.

(2) For the HFPR, Xia and Xu [99] integrated hesitant fuzzy elements in preference relations using different hesitant fuzzy operators, and sorted alternatives according to the aggregation results. Based on the error analysis method, He and Xu [24] presented three priority derivation methods by defining the median, mean value, and deviation value of hesitant fuzzy elements. To consider the consistency property to obtain significant decision-ranking results, a mathematical programming model was introduced to get the ranking through establishing the relationship between a consistent HFPR and its corresponding priorities [134]. After that, by normalizing HFPR and defining the consistency of a HFPR on the basis of complementary judgment matrix, Meng and An [63] constructed a 0–1 mixed programming model to judge the consistency of a HFPR, and further discussed the priority derivation on the assumption that the element of HFPR is independent and uniform with each other. By introducing 0–1 variable to measure the consistency of a HFPR, Tang et al. [82] built a 0–1 mixed programming model to solve the priorities from the perspectives of additive consistency and multiplicative consistency, respectively. Based on the relationship between a consistent HFPR and its corresponding priorities established by Zhu et al. [134] and the concept of "soft consistency" proposed by Kacprzyk and Fedrizzi [36], Xue and Du [115] presented a mathematical programming model for obtaining the priority vector by giving a satisfaction measurement.

(3) There exist priority deviation models for other types of preference relations, such as triangular fuzzy preference relation [17, 26], LPR [34, 67], etc.

At present, the research achievement on the priority deviation models with HFLPR is relatively few. Wang and Gong [89] proposed a model with chance-restricted programming based on measuring hesitation through statistical distributions. Based on defining the worst consistency index for a HFLPR, Chen and Wu [6] put forward an optimization model to get priorities by minimizing the distance between the modified preferences and the original preferences. To improve the applicability of HFLPR, Wu et al. [93, 94, 96] introduced an integer programming model to evaluate the missing value in incomplete one and developed a mixed 0–1 programming model to derive the priority vectors from the complete HFLPRs.

2.2.2 Decision Making with Consistency Measure

Consistency aims to measure whether there is logical contradiction or discrepancy in decision makers' preferences of objects. Although the priority derivation models reviewed in the above section can be used to obtain the ranking of alternatives, it is difficult to guarantee the rationality of the decision making results due to the lack of consistency measure and inconsistency improvement.

For the consistency of preference relation under an uncertain environment, Herrera-Viedma et al. [28] studied the additive consistency of a FPR and proposed a method to construct a consistent FPR based on $n-1$ preference values. Ma et al. [61] introduced two ways based on graph theory to judge the transitivity of FPR based on defining its additive consistency, inconsistency, and weak transitivity. Moreover, they proposed an algorithm to modify the inconsistent FPR. Liu et al. [57] studied the necessary and sufficient conditions for the consistency of a FPR, and developed a method for an incomplete FPR based on least squares. Considering that the existing methods may not guarantee the order consistency, Lee [41] constructed a decision making method based on incomplete FPR from the perspective of additive consistency and order consistency.

Similarly, the research on consistency definition and inconsistency improvement has been generally addressed for other types of vague preference relation, some of which can be listed as follows:

(1) For the IFPR, the consistency measures were discussed, and the procedures for modifying the inconsistent IFPR into an acceptable one were studied [112, 114]. The intuitionistic fuzzy analytic hierarchy process and intuitionistic fuzzy analysis network analysis method were addressed to improve the applicability of IFPR in group decision making [48, 112]. They include the processes of building the hierarchy and network structures of decision making problems, checking consistency of individual IFPRs, repairing the inconsistent IFPRs, and integrating all acceptably consistent IFPRs into an overall one for ranking alternatives. Yang et al. [118] focused on the additively consistent IFPR, captured it with Tanino's normalized priority vector, and demonstrated models to achieve the T-normalization and consistent one. Other discussions on the consistency of an IFPR can refer to the overview given by Ren et al. [76].

(2) For the HFPR, Zhu and Xu [133] studied its multiplicative consistency and provided a completely consistent HFPR. Based on this work, the distance between a HFPR and its corresponding completely consistent HFPR was used to measure consistency, and a decision making algorithm including consistency checking, inconsistency improving and priority deviating was presented [134]. Similarly, Liao et al. [49] studied a decision making algorithm to ensure the consistency of HFPRs from a different consistency measure. Zhang et al. [129] gave a clustering algorithm to transform the inconsistent HFPR into an acceptable one. Liu et al. [54] pointed out that it is unreasonable to set the consistency index of a HFPR as 0.1 without any theoretical support as addressed in [49]. To make up for such a defect, they used Chebyshev inequality to discuss the consistency thresholds for different orders of HFPRs, which provides a different research perspective for the consistency of HFPR and makes the decision making process more rigorous and reasonable. Based on investigating the consistency index and the acceptable consistency, Li and Chen [43] derived some conditions for judging the consistency of HFPR and constructed the characteristic HFPR with satisfying multiplicative consistency. Zhang et al. [125] put forward to repair the inconsistent HFPR into an acceptably consistent one through minimizing the deviation between it and its corresponding perfect one.

Another multiplicative representation of HFPR was addressed, and a mathematical program was subsequently shown to obtain an acceptably consistent one [87].

(3) There exists research on consistency definition, consistency checking, and consistency improvement for other types of preference relations, such as LPR [10–13, 84, 100], 2-tuple LPR [124], triangular fuzzy preference relation [51, 64, 91], etc.

As for the HFLPR, two algorithms to modify the unacceptably consistent one with respect to the additive consistency and multiplicative consistency were developed, respectively [130, 132]. By mapping a HFLPR to an equivalent 2-tuple LPR, Wu and Xu [97] proposed the consistency index of HFLPR. Li et al. [44] came up with an optimization method for measuring the pessimistic consistency index and optimistic consistency index to estimate the consistency range of a HFLPR. Considering that different people may have different understandings of the same linguistic information, Li et al. [45] corresponded the linguistic representation into numerical ranges in personalized decision making with HFLPRs and built an optimization model to maximize the average consistency measure. Furthermore, the properties of additive consistency and multiplicative consistency of HFLPRs were discussed, and the algorithms for improving the two consistency indexes were presented [53]. Li et al. [46] put forward an interval consistency index to manifest the consistency of a HFLPR. Based on a different definition of consistency of a HFLPR, Feng et al. [15] proposed a consistency measurement based on goal programming. They pointed out that the consistency can be judged by LPR's geometric consistency and demonstrated a modeling optimization method and an iterative optimization method to improve the inconsistency. Later on, optimization models were established to modify the inconsistent HFLPR into an acceptable one with multiplicative consistency [95, 96]. Other studies dealing with the consistency of HFLPR can be found in [69, 127, 128].

Generally, the procedure of decision making with HFLPRs based on consistency measure can be shown in Fig. 2.1.

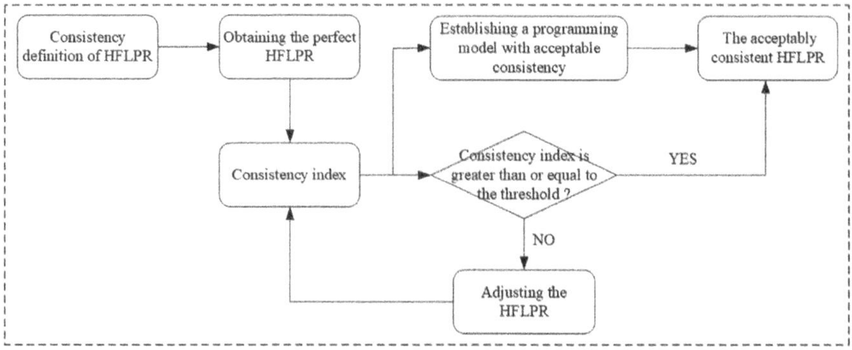

Fig. 2.1 Decision making with HFLPRs based on consistency measure

2.2.3 Consensus Models for Group Decision Making

The complexity and uncertainty of decision making problems determine the broad application of group decision making in real life, which helps make use of group wisdom to ensure the effectiveness of results. It allows decision makers to exchange their decision making opinions and adjust individual opinions that deviate significantly from the group opinions to achieve acceptable results for all decision makers.

Herrera-Viedma et al. [27] established a feedback adjustment model based on the individual consistency and group consensus measurement of incomplete FPRs. Parreiras et al. [73] studied how to standardize the information flow judgments through compatibility index and consistency index and further built a dynamic consensus reaching mechanism. Based on introducing the distance measures, Xu et al. [104] proposed a consensus model for group decision making with FPRs and multiplicative preference relations (MPRs) with defining an individual to group consensus index and a group consensus index. Furthermore, the relevant research for other types of vague preference relation can be briefed as follows:

(1) For the group decision making with IFPRs, the consensus measure was proposed based on the definition of pairwise consistency degree of the IFPRs [81]. Later on, a group decision making algorithm was provided based on computing the compatibility of IFPRs given by any two decision makers [109]. Considering that the consistency of individual judgments is one of the critical factors in scientific decision making, Liao et al. [50] constructed a consensus reaching model based on the consistency of individual IFPRs. Meng et al. [62] established a group consensus process for IFPRs, which includes establishing a 0–1 mixed mathematical model to enhance the individual consistency, the construction of consensus index of IFPRs, and the consistency improvement for the inconsistent IFPRs.

(2) For the group decision making with HFPRs, Liao et al. [49] showed an algorithm for group consensus based on measuring the distance between an individual HFPR and the collective HFPR. Zhang et al. [129] further built a geometric compatibility index to measure the group consensus degree and created an iterative algorithm for achieving group consensus. Other group consensus models based on HFPRs have been established based on discussing different group consensus measurements [82, 102, 105, 126]. These models can ensure the effectiveness and practicability of decision making results since they make the logic of individual HFPR reasonable through consistency driven and avoid the bias caused by lack of knowledge or cognition through group wisdom. In addition, Xu et al. [105] pointed out that the group consensus can be quickly reached with the weighted HFPRs during the decision making process and put forward a group consensus model on the basis of constructing a least-square integrated programming model to synthesize all decision makers' HFPRs. Two programming models were presented to derive the priorities for achieving the expected consensus level and acceptable consistency [47, 87].

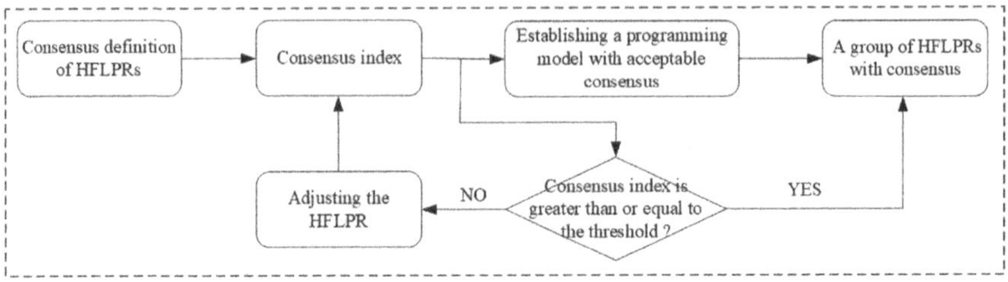

Fig. 2.2 Group consensus with HFLPRs

(3) The group consensus with different types of preference relations have been addressed in the past few decades, such as fuzzy preferences and fuzzy types of majority [37–40], LPR [29, 79, 84, 101], 2-tuple LPR [10, 19, 124].

For the group decision making with HFLPRs, Wu and Xu [97] established a feedback system to handle group consensus based on the similarity matrix of the transformed preference relations. Similarly, through transforming the HFLPR into probabilistic linguistic preference, Liu et al. [56] proposed the projection of the TPR on the overall TPR and then introduced a programming model to derive the priorities with the highest consistency level. Tang and Liao [83] put forward to obtain the rankings of alternatives from the HFLPRs and gave the group consensus process based on the ordinal consensus measure for the obtained ranking. Based on the generalized distance measure of hesitant fuzzy linguistic information, Gou et al. [23] defined the concepts of compatibility degree, compatibility index, and acceptable compatibility for a HFLPR. By calculating the deviations between the individual HFLPRs and the overall HFLPR, different feedback mechanisms were introduced to achieve group consensus [103, 128]. Differently, group consensus reaching models were built with the similarity measurement between an individual HFLPR and other individual HFLPRs [5, 96, 127]. Liu and Jiang [52] further addressed the group consensus model on the basis of the weighted distance between an HFLPR and other HFLPRs.

The procedure of group consensus with HFLPRs can be manifested in Fig. 2.2.

References

1. Abbas, A.E.: Entropy methods for joint distributions in decision analysis. IEEE Trans. Eng. Manage. **53**(1), 146–159 (2006)
2. Buckley, J.J.: Entropy principles in decision making under risk. Risk Anal. **5**(4), 303–313 (1985)
3. Chen, H.Y., Zhou, L.G.: A relative entropy approach to group decision making with interval reciprocal relations based on COWA operator. Group Decis. Negot. **21**, 585–599 (2012)
4. Chen, L., Wang, Y.Z.: Research on TOPSIS integrated evaluation and decision method based on entropy coefficient. Control Decis. **18**(4), 456–459 (2003)

5. Chen, X., Peng, L.J., Wu, Z.B., Pedrycz, W.: Controlling the worst consistency index for hesitant fuzzy linguistic preference relations in consensus optimization models. Comput. Ind. Eng. **143**, 136423 (2020)
6. Chen, X., Wu, Z.B.: An optimization method to improve the worst consistency level of hesitant fuzzy linguistic preference relations. In: 2019 12th International Symposium on Computational Intelligence and Design (ISCID) (2019)
7. Chen, Y., Qu, L.L.: Evaluating the selection of logistics centre location using fuzzy MCDM model based on entropy weight. In: 2006 6th World Congress on Intelligent Control and Automation (2006)
8. Dai, J.H., Wang, W.T., Tian, H.W., Liu, L.: Attribute selection based on a new conditional entropy for incomplete decision systems. Knowl.-Based Syst. **39**, 207–213 (2013)
9. Dai, J.H., Wang, W.T., Xu, Q., Tian, H.W.: Uncertainty measurement for interval-valued decision systems based on extended conditional entropy. Knowl.-Based Syst. **27**, 443–450 (2012)
10. Dong, Y.C., Li, C.C., Herrera, F.: An optimization-based approach to adjusting unbalanced linguistic preference relations to obtain a required consistency level. Inf. Sci. **292**, 27–38 (2015)
11. Dong, Y.C., Li, C.C., Xu, Y.F., Gu, X.: Consensus-based group decision making under multi-granular unbalanced 2-tuple lnguistic preference relations. Group Decis. Negot. **24**, 217–242 (2015)
12. Dong, Y.C., Xu, Y.F.: Consistency measures of linguistic preference relations and its properties in group decision making. In: International Conference on Fuzzy Systems and Knowledge Discovery. Springer-Verlag (2006)
13. Dong, Y.C., Xu, Y.F., Li, H.Y.: On consistency measures of linguistic preference relations. Eur. J. Oper. Res. **189**(2), 430–444 (2008)
14. Fan, Z.P., Ma, J., Jiang, Y.P., Sun, Y.H., Ma, L.: A goal programming approach to group decision making based on multiplicative preference relations and fuzzy preference relations. Eur. J. Oper. Res. **174**(1), 311–321 (2006)
15. Feng, X.Q., Zhang, L., Wei, C.P.: The consistency measures and priority weights of hesitant fuzzy linguistic preference relations. Appl. Soft Comput. **65**, 79–90 (2018)
16. Garg, H., Agarwal, N., Choubey, A.: Entropy based multi-criteria decision making method under fuzzy environment and unknown attribute weights. Glob. J. Technol. Optim. **6**(3), 13–20 (2015)
17. Gong, Y.B.: A method for priority of triangular fuzzy number reciprocal judgment matrix. In: 2007 IEEE International Conference on Grey Systems and Intelligent Services (2007)
18. Gong, Z.W.: Least-square method to priority of the fuzzy preference relations with incomplete information. Int. J. Approximate Reasoning **47**(2), 258–264 (2008)
19. Gong, Z.W., Forrest, J., Yang, Y.J.: the optimal group consensus models for 2-tuple linguistic preference relations. Knowl. Based Syst. **37**(2), 427–437 (2013)
20. Gong, Z.W., Tan, X., Yang, Y.J.: Optimal weighting models based on linear uncertain constraints in intuitionistic fuzzy preference relations. J. Oper. Res. Soc. **70**(8), 1296–1307 (2018)
21. Gong, Z.W., Zhang, N., Chiclana, F.: The optimization ordering model for intuitionistic fuzzy preference relations with utility functions. Knowl.-Based Syst. **162**, 174–184 (2018)
22. Gou, X.J., Xu, Z.S., Liao, H.C.: Hesitant fuzzy linguistic entropy and cross-entropy measures and alternative queuing method for multiple criteria decision making. Inf. Sci. **388–389**, 225–246 (2017)
23. Gou, X.J., Xu, Z.S., Liao, H.C.: Group decision making with compatibility measures of hesitant fuzzy linguistic preference relations. Soft. Comput. **23**, 1511–1527 (2019)
24. He, Y., Xu, Z.S.: Error analysis methods for group decision making based on hesitant fuzzy preference relation. Int. J. Intell. Syst. **31**, 1104–1128 (2016)
25. He, Y.H., Guo, H.M., Jin, M.Z., Ren, P.Y.: A linguistic entropy weight method and its application in linguistic multi-attribute group decision making. Nonlinear Dyn. **84**(1), 399–404 (2016)

26. He, Y.Y., Zhou, D.Q., Wang, Q.: Least variance priority method for triangular fuzzy number complementary judgment matrix. Control Decis. **23**(10), 1113–1116 (2008)
27. Herrera-Viedma, E., Alonso, S., Chiclana, F., Herrera, F.: A consensus model for group decision making with incomplete fuzzy preference relations. IEEE Trans. Fuzzy Syst. **15**(5), 863–877 (2007)
28. Herrera-Viedma, E., Herrera, F., Chiclana, F., Luque, M.: Some issues on consistency of fuzzy preference relations. Eur. J. Oper. Res. **154**(1), 98–109 (2004)
29. Herrera-Viedma, E., Martinez, L., Mata, F., Chiclana, F.: A consensus support system model for group decision making problems with multigranular linguistic preference relations. IEEE Trans. Fuzzy Syst. **13**(5), 644–658 (2005)
30. Hsu, L.C.: Investment decision making using a combined factor analysis and entropy-based topsis model. J. Bus. Econ. Manag. **14**(3), 448–466 (2013)
31. Hu, Q.H., Che, X.J., Zhang, L., Zhang, D., Guo, M.Z., Yu, D.R.: Rank entropy-based decision trees for monotonic classification. IEEE Trans. Knowl. Data Eng. **24**(11), 2052–2064 (2012)
32. Ichihashi, H., Shirai, T., Nagasaka, K., Miyosh, T.: Neuro-fuzzy ID3: a method of inducing fuzzy decision trees with linear programming for maximizing entropy and an algebraic method for incremental learning. Fuzzy Sets Syst. **81**, 157–167 (1996)
33. Jiang, F., Sui, Y.F., Zhou, L.: A relative decision entropy-based feature selection approach. Pattern Recogn. **48**(7), 2151–2163 (2015)
34. Jiang, J., Li, X., Yang, K.W., Chen, Y.W.: A goal programming approach to Group Decision Making with incomplete linguistic preference relations. In: IEEE International Conference on Intelligent Computing and Intelligent Systems (2010)
35. Jin, F.F., Pei, L.D., Chen, H.Y., Zhou, L.G.: Interval-valued intuitionistic fuzzy continuous weighted entropy and its application to multi-criteria fuzzy group decision making. Knowl.-Based Syst. **59**, 132–141 (2014)
36. Kacprzyk, J., Fedrizzi, M.: A "soft" measure of consensus in the setting of partial (fuzzy) preferences. Eur. J. Oper. Res. **34**(3), 316–325 (1988)
37. Kacprzyk, J., Fedrizzi, M., Nurmi, H.: Group decision making and consensus under fuzzy preferences and fuzzy majority. Fuzzy Sets Syst. **49**(1), 21–31 (1992)
38. Kacprzyk, J., Fedrizzi, M., Nurmi, H.: OWA operators in group decision making and consensus reaching under fuzzy preferences and fuzzy majority. In: The Ordered Weighted Averaging Operators, pp. 193–206 (1997).
39. Kacprzyk, J., Nurmi, H., Fedrizzi, M.: Group decision making and a measure of consensus under fuzzy preferences and a fuzzy linguistic majority. Comput. Words Inf./Intell. Syst. 2 Stud. Fuzziness Soft Comput. **34**, 243–269 (1999)
40. Kacprzyk, J., Zadrozny, S., Fedrizzi, M., Nurmi, H.: On group decision making, consensus reaching, voting and voting paradoxes under fuzzy preferences and a fuzzy majority: a survey and some perspectives. In: Studies in Fuzziness and Soft Computing, vol. 220, (2008)
41. Lee, L.W.: Group decision making with incomplete fuzzy preference relations based on the additive consistency and the order consistency. Expert Syst. Appl. **39**(14), 11666–11676 (2012)
42. Lee, M.Y., Yang, C.S.: Entropy-based feature extraction and decision tree induction for breast cancer diagnosis with standardized thermograph images. Comput. Methods Programs Biomed. **100**(3), 269–282 (2010)
43. Li, B.P., Chen, Y.H.: Determination conditions on multiplicative consistency of hesitant fuzzy preference relations and their applications to group decision making. Fuzzy Syst. Math. **30**(5), 157–168 (2016)
44. Li, C.C., Dong, Y.C., Herrera, F., Martínez, L.: An optimization-based approach to estimate the range of consistency in hesitant fuzzy linguistic preference relations. In: 2016 IEEE International Conference on Fuzzy Systems (2016)
45. Li, C.C., Rodríguez, R.M., Herrera, F., Martinez, L., Dong, Y.C.: A consistency-driven approach to set personalized numerical scales for hesitant fuzzy linguistic preference relations. In: 2017 IEEE International Conference on Fuzzy Systems (2017)

46. Li, C.C., Rodríguez, R.M., Martínez, L., Dong, Y.C., Herrera, F.: Consistency of hesitant fuzzy linguistic preference relations: An interval consistency index. Inf. Sci. **432**, 347–361 (2018)
47. Li, J., Wang, Z.X.: Deriving priority weights from hesitant fuzzy preference relations in view of additive consistency and consensus. Soft Comput. **23**(2), 13691–13707 (2019)
48. Liao, H.C., Mi, X.M., Xu, Z.S., Xu, J.P., Herrera, F.: Intuitionistic fuzzy analytic network process. IEEE Trans. Fuzzy Syst. **26**(5), 2578–2590 (2018)
49. Liao, H.C., Xu, Z.S., Xia, M.M.: Multiplicative consistency of hesitant fuzzy preference relation and its application in group decision making. Int. J. Inf. Technol. Decis. Making **13**(01), 47–76 (2014)
50. Liao, H.C., Xu, Z.S., Zeng, X.J., Merigó, J.M.: Framework of group decision making with intuitionistic fuzzy preference information. IEEE Trans. Fuzzy Syst. **23**(4), 1211–1227 (2015)
51. Liu, F., Zhang, W.G., Zhang, L.H.: Consistency analysis of triangular fuzzy reciprocal preference relations. Eur. J. Oper. Res. **235**(3), 718–726 (2014)
52. Liu, H.B., Jiang, J.: Optimizing consistency and consensus improvement process for hesitant fuzzy linguistic preference relations and the application in group decision making. Inf. Fusion **56**, 114–127 (2020)
53. Liu, H.B., Jiang, L., Xu, Z.S.: Improving the additive and multiplicative consistency of hesitant fuzzy linguistic preference relations. J. Intell. Fuzzy Syst. **33**(6), 3677–3693 (2017)
54. Liu, H.F., Xu, Z.S., Liao, H.C.: The multiplicative consistency index of hesitant fuzzy preference relation. IEEE Trans. Fuzzy Syst. **24**(1), 82–93 (2016)
55. Liu, M.F., Ren, H.P.: A new intuitionistic fuzzy entropy and application in multi-attribute decision making. Information **5**(4), 587–601 (2014)
56. Liu, N.N., He, Y., Xu, Z.S.: A new approach to deal with consistency and consensus issues for hesitant fuzzy linguistic preference relations. Appl. Soft Comput. **76**, 400–415 (2019)
57. Liu, X.W., Pan, Y.W., Xu, Y.J., Yu, S.: Least square completion and inconsistency repair methods for additively consistent fuzzy preference relations. Fuzzy Sets Syst. **198**, 1–19 (2012)
58. Lotfi, F.H., Fallahnejad, R.: Imprecise Shannon's entropy and multi attribute decision making. Entropy **12**, 53–62 (2010)
59. Luo, D., Liu, S.F.: Analytic method to a kind of grey fuzzy decision making based on entropy. Eng. Sci. **6**(10), 48–51 (2004)
60. Luo, Y., Li, Y.L.: Comprehensive decision making of transmission network planning based on entropy weight and grey relational analysis. Power Syst. Technol. **37**(1), 77–81 (2013)
61. Ma, J., Fan, Z.P., Jiang, Y.P., Mao, J.Y., Ma, L.: A method for repairing the inconsistency of fuzzy preference relations. Fuzzy Sets Syst. **157**(1), 20–33 (2006)
62. Meng, F.Y., Tang, J., Xu, Z.S.: A 0–1 mixed programming model based method for group decision making with intuitionistic fuzzy preference relations. Comput. Ind. Eng. **112**, 289–304 (2017)
63. Meng, F.Y., An, Q.X.: A new approach for group decision making method with hesitant fuzzy preference relations. Knowl.-Based Syst. **127**, 1–15 (2017)
64. Meng, F.Y., Chen, X.H.: A robust additive consistency-based method for decision making with triangular fuzzy reciprocal preference relations. Fuzzy Optim. Decis. Making **17**, 49–73 (2018)
65. Meng, F.Y., Chen, X.H., Zhang, Y.L.: Consistency-based linear programming models for generating the priority vector from interval fuzzy preference relations. Appl. Soft Comput. **41**, 247–264 (2016)
66. Meng, F.Y., Tang, J.: Interval-valued intuitionistic fuzzy multiattribute group decision making based on cross entropy measure and Choquet integral. Int. J. Intell. Syst. **28**(12), 1172–1195 (2013)
67. Meng, F.Y., Tang, J., Xu, Z.S.: Exploiting the priority weights from interval linguistic fuzzy preference relations. Soft. Comput. **23**, 583–597 (2019)
68. Mi, C.M., Liu, S.F., Dang, Y.G., Zhou, J.: Study on grey fixed weight clustering decision making based on entropy weight. In: 2006 IEEE International Conference on Systems, Man, and Cybernetics (2006)

69. Mi, X.M., Wu, X.L., Tang, M., Liao, H.C., Al-Barakati, A., Altalhi, A.H., Herrera, F.: Hesitant fuzzy linguistic analytic hierarchical process with prioritization, consistency checking, and inconsistency repairing. IEEE Access **7**, 44135–44149 (2019)
70. Mon, D.L., Cheng, C.H., Lin, J.C.: Evaluating weapon system using fuzzy analytic hierarchy process based on entropy weight. Fuzzy Sets Syst. **62**(2), 127–134 (1994)
71. Montesarchio, V., Ridolfi, E., Russo, F., Napolitano, F.: Rainfall threshold definition using an entropy decision approach and radar data. Nat. Hazard. **11**(7), 2061–2074 (2011)
72. Myung, I.J.: Maximum entropy interpretation of decision bound and context models of categorization. J. Math. Psychol. **38**(3), 335–365 (1994)
73. Parreiras, R., Ekel, P., Bernardes, F., Jr.: A dynamic consensus scheme based on a nonreciprocal fuzzy preference relation modeling. Inf. Sci. **21**, 11–17 (2012)
74. Qi, X.W., Liang, C.Y., Zhang, E.Q., Ding, Y.: Approach to interval-valued intuitionistic fuzzy multiple attributes group decision making based on maximum entropy. Syst. Eng. Theory Pract. **31**(10), 1940–1948 (2011)
75. Qi, X.W., Liang, C.Y., Zhang, J.L.: Generalized cross-entropy based group decision making with unknown expert and attribute weights under interval-valued intuitionistic fuzzy environment. Comput. Ind. Eng. **79**, 52–64 (2015)
76. Ren, P.J., Xu, Z.S., Kacprzyk, J.: Group decisions with intuitionistic fuzzy sets. Springer Handbook of Group Decision and Negotiation , accepted (2020)
77. Szmidt, E., Kacprzyk, J.: Entropy for intuitionistic fuzzy sets. Fuzzy Sets Syst. **118**, 467–477 (2001)
78. Smith, S.A.: A derivation of entropy and the maximum entropy criterion in the context of decision problems. In: IEEE Transactions on Systems, Man, and Cybernetics—Part A: Systems and Humans, SMC-4157–163 (1974)
79. Sun, B.Z., Ma, W.M.: An approach to consensus measurement of linguistic preference relations in multi-attribute group decision making and application. Omega **51**, 83–92 (2015)
80. Sun, L., Xu, J.C., Tian, Y.: Feature selection using rough entropy-based uncertainty measures in incomplete decision systems. Knowl.-Based Syst. **36**, 206–216 (2012)
81. Szmidt, E., Kacprzyk, J.: A consensus-reaching process under intuitionistic fuzzy preference relations. Int. J. Intell. Syst. **18**, 837–852 (2003)
82. Tang, J., An, Q.X., Meng, F.Y., Chen, X.H.: A natural method for ranking objects from hesitant fuzzy preference relations. Int. J. Inf. Technol. Decis. Mak. **16**(06), 1611–1646 (2017)
83. Tang, M., Liao, H.C.: Multiple criteria group decision making based on hesitant fuzzy linguistic consensus model for fashion sales forecasting. In: International Conference on Artificial Intelligence on Textile and Apparel (2018)
84. Tian, J.F., Zhang, Z., Ha, M.H.: An additive consistency and consensus-based approach for uncertain group decision making with linguistic preference relations. IEEE Trans. Fuzzy Syst. **27**(5), 873–887 (2019)
85. Tian, Z.P., Zhang, H.Y., Wang, J., Wang, J.Q., Chen, X.H.: Multi-criteria decision making method based on a cross-entropy with interval neutrosophic sets. Int. J. Syst. Sci. **47**(15), 3598–3608 (2016)
86. Verma, M., Rajasankar, J.: A thermodynamical approach towards group multi-criteria decision making (GMCDM) and its application to human resource selection. Appl. Soft Comput. **52**, 323–332 (2017)
87. Wan, S.P., Zhong, L.G., Dong, J.Y.: A new method for group decision making with hesitant fuzzy preference relations based on multiplicative consistency. IEEE Trans. Fuzzy Syst. **28**(7), 1449–1463 (2020)
88. Wang, J., Lan, J.B., Ren, P.Y., Luo, Y.Y.: Some programming models to derive priority weights from additive interval fuzzy preference relation. Knowl.-Based Syst. **27**(3), 69–77 (2012)
89. Wang, L.H., Gong, Z.W.: Priority of a hesitant fuzzy linguistic preference relation with a normal distribution in meteorological disaster risk assessment. Int. J. Environ. Res. Public Health **14**(10), 1203 (2017)
90. Wang, Y.M., Fan, Z.P., Hua, Z.S.: A chi-square method for obtaining a priority vector from multiplicative and fuzzy preference relations. Eur. J. Oper. Res. **182**(1), 356–366 (2007)

91. Wang, Z.J.: Consistency analysis and priority derivation of triangular fuzzy preference relations based on modal value and geometric mean. Inf. Sci. **314**, 169–183 (2015)
92. Wu, J.Z., Zhang, Q.: Multicriteria decision making method based on intuitionistic fuzzy weighted entropy. Expert Syst. Appl. **38**, 916–922 (2011)
93. Wu, P., Li, H.Y., Merigó, J.M., Zhou, L.G.: Integer programming modeling on group decision making with incomplete hesitant fuzzy linguistic preference relations. IEEE Access **7**, 136867–136881 (2019)
94. Wu, P., Zhou, L.G., Chen, H.Y., Tao, Z.F.: Additive consistency of hesitant fuzzy linguistic preference relation with a new expansion principle for hesitant fuzzy linguistic term sets. IEEE Trans. Fuzzy Syst. **27**(4), 716–730 (2019)
95. Wu, P., Zhou, L.G., Chen, H.Y., Tao, Z.F.: Multi-stage optimization model for hesitant qualitative decision making with hesitant fuzzy linguistic preference relations. Appl. Intell. **50**, 222–240 (2020)
96. Wu, P., Zhu, J., Zhou, L., Chen, H.: Local feedback mechanism based on consistency-derived for consensus building in group decision making with hesitant fuzzy linguistic preference relations. Comput. Ind. Eng. **137**, 106001 (2019)
97. Wu, Z.B., Xu, J.P.: Managing consistency and consensus in group decision making with hesitant fuzzy linguistic preference relations. Omega **65**, 28–40 (2016)
98. Xia, M.M., Xu, Z.S.: Entropy/cross entropy-based group decision making under intuitionistic fuzzy environment. Inf. Fusion **13**(1), 31–47 (2012)
99. Xia, M.M., Xu, Z.S.: Managing hesitant information in GDM problems under fuzzy and multiplicative preference relations. Int. J. Uncertainty Fuzziness Knowl. Based Syst. **21**(06), 865–897 (2013)
100. Xia, M.M., Xu, Z.S., Wang, Z.: Multiplicative consistency-based decision support system for incomplete linguistic preference relations. Int. J. Syst. Sci. **45**(3), 625–636 (2014)
101. Xu, J.P., Wu, Z.B.: A maximizing consensus approach for alternative selection based on uncertain linguistic preference relations. Comput. Ind. Eng. **64**(4), 999–1008 (2013)
102. Xu, X.Y., Cabrerizo, F.J., Herrera-Viedma, E.: A consensus model for hesitant fuzzy preference relations and its application in water allocation management. Appl. Soft Comput. **58**, 265–284 (2017)
103. Xu, Y., Wen, X., Sun, H., Wang, H.M.: Consistency and consensus models with local adjustment strategy for hesitant fuzzy linguistic preference relations. Int. J. Fuzzy Syst. **20**(7), 2216–2233 (2018)
104. Xu, Y.J., Li, K.W., Wang, H.M.: Distance-based consensus models for fuzzy and multiplicative preference relations. Inf. Sci. **253**, 56–73 (2013)
105. Xu, Y.J., Rui, D., Wang, H.M.: A dynamically weight adjustment in the consensus reaching process for group decision making with hesitant fuzzy preference relations. Int. J. Syst. Sci. **48**(6), 1311–1321 (2017)
106. Xu, Z.S.: Approaches to multiple attribute decision making with intuitionistic fuzzy preference information. Syst. Eng. Theory Pract. **27**(11), 62–71 (2007)
107. Xu, Z.S.: Intuitionistic preference relations and their application in group decision making. Inf. Sci. **177**, 2363–2379 (2007)
108. Xu, Z.S.: An error-analysis-based method for the priority of an intuitionistic preference relation in decision making. Knowl.-Based Syst. **33**, 173–179 (2012)
109. Xu, Z.S.: Compatibility analysis of intuitionistic fuzzy preference relations in group decision making. Group Decis. Negot. **22**(3), 463–482 (2013)
110. Xu, Z.S., Da, Q.L.: A least deviation method to obtain a priority vector of a fuzzy preference relation. Eur. J. Oper. Res. **164**(1), 206–216 (2005)
111. Xu, Z.S., Hu, H.: Entropy-based procedures for intuitionistic fuzzy multiple attribute decision making. J. Syst. Eng. Electron. **20**(5), 1001–1011 (2009)
112. Xu, Z.S., Liao, H.C.: Intuitionistic fuzzy analytic hierarchy process. IEEE Trans. Fuzzy Syst. **22**, 749–761 (2014)
113. Xu, Z.S., Xia, M.M.: Hesitant fuzzy entropy and cross-entropy and their use in multiattribute decision making. Int. J. Intell. Syst. **27**(9), 799–822 (2012)

114. Xu, Z.S., Xia, M.M.: Iterative algorithms for improving consistency of intuitionistic preference relations. J. Oper. Res. Soc. **65**(5), 708–722 (2014)
115. Xue, M., Du, Y.F.: A group decision making model based on regression method with hesitant fuzzy preference relations. Math. Probl. Eng. **2017**, 1–8 (2017)
116. Yager, R.R.: Weighted maximum entropy OWA aggregation with applications to decision making under risk. IEEE Trans. Syst. Man Cybernet. Part A: Syst. Humans **39**(3), 555–564 (2009)
117. Yang, J.P., Qiu, W.H.: A measure of risk and a decision making model based on expected utility and entropy. Eur. J. Oper. Res. **164**, 792–799 (2005)
118. Yang, W., Jhang, S.T., Shi, S.G., Xu, Z.S., Ma, Z.M.: A novel additive consistency for intuitionistic fuzzy preference relations in group decision making. Appl. Intell. **50**, 4342–4356 (2020)
119. Ye, J.: Multicriteria fuzzy decision making method based on the intuitionistic fuzzy cross-entropy. International Conference on Intelligent Human-Machine Systems and Cybernetics (2009)
120. Ye, J.: Multicriteria fuzzy decision making method using entropy weights-based correlation coefficients of interval-valued intuitionistic fuzzy sets. Appl. Math. Model. **34**(12), 3864–3870 (2010)
121. Ye, J.: Fuzzy cross entropy of interval-valued intuitionistic fuzzy sets and its optimal decision making method based on the weights of alternatives. Expert Syst. Appl. **38**, 6179–6183 (2011)
122. Ye, J.: Single valued neutrosophic cross-entropy for multicriteria decision making problems. Appl. Math. Model. **38**(3), 1170–1175 (2014)
123. Zhang, Y.J., Li, P.H., Wang, Y.Z., Ma, P.J., Su, X.H.: Multiattribute decision making based on entropy under interval-valued intuitionistic fuzzy environment. Math. Prob. Eng. 526871 (2013)
124. Zhang, Z., Guo, C.H.: Consistency and consensus models for group decision making with uncertain 2-tuple linguistic preference relations. Int. J. Syst. Sci. **47**, 2572–2587 (2016)
125. Zhang, Z., Kou, X.Y., Dong, Q.X.: Additive consistency analysis and improvement for hesitant fuzzy preference relations. Expert Syst. Appl. **98**, 118–128 (2018)
126. Zhang, Z.M.: A framework of group decision making with hesitant fuzzy preference relations based on multiplicative consistency. Int. J. Fuzzy Syst. **19**(4), 982–996 (2017)
127. Zhang, Z.M., Chen, S.M.: Group decision making based on acceptable multiplicative consistency and consensus of hesitant fuzzy linguistic preference relations. Inf. Sci. **541**, 531–550 (2020)
128. Zhang, Z.M., Chen, S.M.: Group decision making with hesitant fuzzy linguistic preference relations. Inf. Sci. **514**, 354–368 (2020)
129. Zhang, Z.M., Wang, C., Tian, X.D.: A decision support model for group decision making with hesitant fuzzy preference relations. Knowl.-Based Syst. **86**, 77–101 (2015)
130. Zhang, Z.M., Wu, C.: On the use of multiplicative consistency in hesitant fuzzy linguistic preference relations. Knowl.-Based Syst. **72**, 13–27 (2014)
131. Zhao, M., Qiu, W.H., Liu, B.S.: Relative entropy evaluation method for multiple attribute decision making. Control Decis. **25**(7), 1098–1104 (2010)
132. Zhu, B., Xu, Z.S.: Consistency measures for hesitant fuzzy linguistic preference relations. IEEE Trans. Fuzzy Syst. **22**, 34–45 (2014)
133. Zhu, B., Xu, Z.S.: Regression methods for hesitant fuzzy preference relations. Technol. Econ. Dev. Econ. **19**(1), S214–S227 (2014)
134. Zhu, B., Xu, Z.S., Xu, J.P.: Deriving a ranking from hesitant fuzzy preference relations under group decision making. IEEE Trans. Cybernet. **44**(8), 1328–1337 (2014)

Chapter 3
A Thermodynamic Method for Intuitionistic Fuzzy Decision Making

As the IFS [1, 2] is a more general concept of Zadeh's fuzzy set to portray the uncertainty of things effectively, the decision-making methods with IFSs, such as TOPSIS [3], VIKOR (Vise Kriterijumska Optimizacija Kompromisno Resenje) [7], ELECTRE (Elimination and Choice Expressing Reality) [14], LINMAP (Linear Programming Techniques for Multidimensional Analysis of Preference) [5, 6, 9], AHP (Analytic Hierarchy Process) [17], WASPAS (Weighted Aggregated Sum Product Assessment) [20], have been proposed. However, they only use the numerical performance of IFSs to obtain the decision-making results but ignore a vital character, i.e., the distribution of the IFSs. To improve the reliability of decision making results based on considering both numerical performance and distribution of IFSs, in what follows, we introduce the thermodynamic method for decision making with IFS studied in [10].

3.1 Intuitionistic Fuzzy Set

Definition 3.1 [2]. Let $X = \{x_1, x_2, \ldots, x_n\}$ be a fixed set. Then an IFS A on X is

$$A = \{\langle x, \mu_A(x), \nu_A(x)\rangle | x \in X\} \tag{3.1}$$

where $\mu_A(x)$ and $\nu_A(x)$ are the membership degree and the non-membership degree of x in A, and satisfy $\mu_A(x), \nu_A(x) \geq 0, 0 \leq \mu_A(x) + \nu_A(x) \leq 1, \forall x \in X$. The hesitancy or indeterminacy degree of x in A is $\pi_A(x) = 1 - \mu_A(x) - \nu_A(x)$ [11]. The pair $(\mu_A(x), \nu_A(x))$ is called an intuitionistic fuzzy number (IFN), denoted as $\alpha = (\mu_\alpha, \nu_\alpha)$, where $\mu_\alpha, \nu_\alpha \geq 0$, $\mu_\alpha + \nu_\alpha \leq 1$ [19].

Based on the concepts, the operational laws for IFNs were proposed as follows:

P. Ren and Z. Xu, *Decision-Making Analyses with Thermodynamic Parameters and Hesitant Fuzzy Linguistic Preference Relations*, Studies in Fuzziness and Soft Computing 409, https://doi.org/10.1007/978-3-030-73253-0_3

Definition 3.2 [8, 13, 18]. Let $\alpha = (\mu_\alpha, \nu_\alpha)$, $\alpha_1 = (\mu_{\alpha_1}, \nu_{\alpha_1})$ and $\alpha_2 = (\mu_{\alpha_2}, \nu_{\alpha_2})$ be three IFNs, then

(1) $\alpha^c = (\nu_\alpha, \mu_\alpha)$;
(2) $\alpha_1 \oplus \alpha_2 = (\mu_{\alpha_1} + \mu_{\alpha_2} - \mu_{\alpha_1}\mu_{\alpha_2}, \nu_{\alpha_1}\nu_{\alpha_2})$;
(3) $\alpha_1 \ominus \alpha_2 = (\mu_{\alpha_1\ominus\alpha_2}, \nu_{\alpha_1\ominus\alpha_2}$, where

$$\mu_{\alpha_1\ominus\alpha_2} = \begin{cases} & if\,\mu_1 \geq \mu_2 \; and \; \nu_1 \leq \nu_2 \\ \frac{\mu_1-\mu_2}{1-\mu_2}, & and \; \nu_2 > 0 \\ & and \; \nu_1\pi_2 \leq \pi_1\nu_2 \\ 0, & otherwise \end{cases}$$

$$\nu_{\upsilon_{\alpha_1\ominus\alpha_2}} = \begin{cases} & if\,\mu_1 \geq \mu_2 \; and \; \nu_1 \leq \nu_2 \\ \frac{\nu_1}{\nu_2}, & and \; \nu_2 > 0 \\ & and \; \nu_1\pi_2 \leq \pi_1\nu_2 \\ 1, & otherwise \end{cases}$$

where π_1 and π_2 are the indeterminacy degree of α_1 and α_2;

(4) $\lambda\alpha = (1 - (1 - \mu_\alpha)^\lambda, \nu_\alpha^\lambda), \lambda > 0$.

Later on, Xu and Yager [18] defined the score function and the accuracy function of an IFN $\alpha = (\mu_\alpha, \nu_\alpha)$ as $s(\alpha) = \mu_\alpha - \nu_\alpha$ and $h(\alpha) = \mu_\alpha + \nu_\alpha$, respectively. Then for two IFNs α_1 and α_2:

(1) If $s(\alpha_1) > s(\alpha_2)$, then $\alpha_1 > \alpha_2$;
(2) If $s(\alpha_1) = s(\alpha_2)$, then

  (a) If $h(\alpha_1) > h(\alpha_2)$, then $\alpha_1 > \alpha_2$;
  (b) If $h(\alpha_1) = h(\alpha_2)$, then $\alpha_1 = \alpha_2$.

The Hamming distance [4] between IFNs α_1 and α_2 can be calculated as follows:

$$d(\alpha_1, \alpha_2) = \frac{1}{2}(|\mu_{\alpha_1} - \mu_{\alpha_2}| + |\nu_{\alpha_1} - \nu_{\alpha_2}| + |\pi_{\alpha_1} - \pi_{\alpha_2}|) \tag{3.2}$$

For a collection of IFNs $\alpha_j = (\mu_{\alpha_j}, \nu_{\alpha_j})$ for $j = 1, 2, \ldots, n$, suppose that $w = (w_1, w_2, \ldots, w_n)^T$ is the weight vector of them with $0 \leq w_j \leq 1$ $(j = 1, 2, \ldots, n)$, and $\sum_{j=1}^n w_j = 1$, then the function *IFWA* is called the intuitionistic fuzzy weighted averaging (IFWA) operator if [15]

$$\begin{aligned} IFWA_w(\alpha_1, \alpha_2, \ldots, \alpha_n) &= \mathop{\oplus}_{j=1}^{n} w_j\alpha_j = w_1\alpha_1 \oplus w_2\alpha_2 \oplus \cdots \oplus w_n\alpha_n \\ &= \left(1 - \prod_{j=1}^n (1 - \mu_{\alpha_j})^{\omega_j}, \prod_{j=1}^n \nu_{\alpha_j}^{\omega_j}\right) \end{aligned} \tag{3.3}$$

3.2 Description of the Method with Thermodynamic Parameters

Let a multi-criteria decision making (MCDM) problem contain m alternatives $A = \{A_i | i = 1, 2, \ldots, m\}$ and n criteria $C = \left\{C_j | j = 1, 2, \ldots., n\right\}$. Suppose that h decision-makers $D = \{D_k | k = 1, 2, \ldots. h\}$ assign the weights of criteria as $w = \left\{w_j^k | j = 1, 2, \ldots, n, k = 1, 2, \ldots, h\right\}$ s, and give their evaluations on alternatives with respect to each criterion, denoted as $R = (r_{ij}^k)_{m \times n}$, where r_{ij}^k is a real number for $i = 1, 2, \ldots, m,\ j = 1, 2, \ldots, n,\ k = 1, 2, \ldots, h$. To effectively handle this problem, then the method with thermodynamic parameters was proposed as follows [12]:

(1) Define the decision value r_{ij}^k as the potential of the alternative A_i under the criterion C_j given by kth decision-maker, and define the weight w_j^k as the force correspondingly;

(2) Compute the quality of the potential r_{ij}^k by

$$q_{ij}^k = 1 - \frac{\left| r_{ij}^k - \frac{1}{h}\sum_{k=1}^h r_{ij}^k \right|}{\frac{1}{h}\sum_{k=1}^h r_{ij}^k} \tag{3.4}$$

for $i = 1, 2, \ldots, m,\ j = 1, 2, \ldots, n,\ k = 1, 2, \ldots, h$;

(3) Construct the energy matrices $U^k = (u_{ij}^k)_{m \times n}$ and exergy matrices $X^k = (x_{ij}^k)_{m \times n}$ for $k = 1, 2, \ldots, h$, where

$$u_{ij}^k = w_j^k \cdot r_{ij}^k \tag{3.5}$$

$$x_{ij}^k = q_j^k \cdot u_{ij}^k \tag{3.6}$$

for $i = 1, 2, \ldots, m,\ j = 1, 2, \ldots, n,\ k = 1, 2, \ldots, h$;

(4) Determine the energy index and exergy index respectively by the following formulas:

$$\text{Energy index: } u_i^k = \frac{1}{n}\sum_{j=1}^n u_{ij}^k \tag{3.7}$$

$$\text{Exergy index: } x_i^k = \frac{1}{n}\sum_{j=1}^n x_{ij}^k \tag{3.8}$$

(5) Calculate the overall energy index and the overall exergy index of the alternative A_i as:

$$\text{Overall energy index: } u_i = \frac{1}{h}\sum_{k=1}^{h} u_i^k \tag{3.9}$$

$$\text{Overall exergy index: } x_i = \frac{1}{h}\sum_{k=1}^{h} x_i^k \tag{3.10}$$

(6) Obtain the overall entropy index of the alternative A_i by

$$\text{Overall entropy index: } s_i = u_i - x_i \tag{3.11}$$

rank alternatives according to their overall entropy indexes, the smaller value of the overall entropy index indicates the better performance of the alternative.

3.3 Intuitionistic Fuzzy Decision-Making Method with Thermodynamic Parameters

Based on the method proposed in [12, Ren et al. 10] furtherly investigated the thermodynamic decision-making method with intuitionistic fuzzy information.

For the MCDM problem described in Sect. 3.2, suppose that h decision-makers D_k $(k = 1, 2, \ldots, h)$ provide their evaluations over the alternative A_i with respect to the criterion C_j by the IFNs $\alpha_{ij}^k = (\mu_{ij}^k, \nu_{ij}^k)$, where μ_{ij}^k and ν_{ij}^k are the degrees that alternative A_i satisfies and dissatisfies criterion C_j, respectively, for $i = 1, 2, \ldots, m$, $j = 1, 2, \ldots, n$, $k = 1, 2, \ldots, h$, then the intuitionistic fuzzy decision matrices (IFDMs) $R^k = (r_{ij}^k)_{m\times n}$ $(k = 1, 2, \ldots, h)$ are:

$$R^k = \left(r_{ij}^k\right)_{m\times n} = \begin{array}{c} \\ A_1 \\ A_2 \\ \vdots \\ A_m \end{array}\begin{array}{c} \begin{array}{cccc} C_1 & C_2 & \cdots & C_n \end{array} \\ \begin{pmatrix} (\mu_{11}^k, \nu_{11}^k) & (\mu_{12}^k, \nu_{12}^k) & \cdots & (\mu_{1n}^k, \nu_{1n}^k) \\ (\mu_{21}^k, \nu_{21}^k) & (\mu_{22}^k, \nu_{22}^k) & \cdots & (\mu_{2n}^k, \nu_{2n}^k) \\ \vdots & \vdots & \vdots & \vdots \\ (\mu_{m1}^k, \nu_{m1}^k) & (\mu_{m2}^k, \nu_{m2}^k) & \cdots & (\mu_{mn}^k, \nu_{mn}^k) \end{pmatrix} \end{array}, \; k = 1, 2, \ldots, h \tag{3.12}$$

Based on the above descriptions, Ren et al. [10] gave definitions about thermodynamic decision-making parameters under the intuitionistic fuzzy environment as follows:

Definition 3.3 [10]. Intuitionistic decision potential indicates the potential energy of an alternative with respect to a criterion, which is expressed by the corresponding decision value shown as an IFN.

Definition 3.4 [10]. Intuitionistic decision force is defined as the potential of an alternative with respect to a criterion, which is denoted as the criterion weight.

Ren et al. [10] presented an example to illustrate the above concepts as: if a decision-maker assesses an alternative with respect to a criterion as $\alpha = (0.5, 0.2)$, and the weight of the criterion is 0.35, then the intuitionistic decision potential and the intuitionistic decision force of the object are (0.5, 0.2) and 0.35, respectively.

The intuitionistic decision energy of an alternative was furtherly provided:

Definition 3.5 [10]. In a state, intuitionistic decision energy is an alternative property, which expresses the energy that the alternative possesses in the system. It relates to the intuitionistic decision potential and the intuitionistic decision force, and can be calculated by

$$E_{ij}^k = w_j^k r_{ij}^k \tag{3.13}$$

for $i \in \{1, 2, \ldots, m\}$, $j \in \{1, 2, \ldots, n\}$, $k \in \{1, 2, \ldots, h\}$.

Continuing with the example, the intuitionistic decision energy of the alternative with respect to a criterion is obtained as $E = 0.35 \otimes (0.5, 0.2) = (0.2154, 0.5693)$ [10].

Ren et al. [10] proposed that based on the above definitions, lots of methods can be used to address the intuitionistic decision-making problems, such as the method with aggregations [15, 18], the methods with the TOPSIS idea [3], etc., which demonstrate the results by the numerical value of intuitionistic decision energy. However, Ren et al. [10] stated that the mentioned methods ignore a significant character that is the distribution of the intuitionistic decision potentials. For example, if four decision-makers assess the object 1 as $r_1 = (0.6, 0.6)$, $r_2 = (0.6, 0.6)$, $r_3 = (0.6, 0.4)$ and $r_4 = (0.6, 0.9)$, and the object 2 as $r_1' = (0.6, 0.9)$, $r_2' = (0.6, 0.4)$, $r_3' = (0.6, 0.4)$ and $r_4' = (0.6, 0.9)$, then by the IFWA operator under the assumptions of the same weight for the decision-makers, the intuitionistic decision energy of the object 1 is equal to that of the object 2. Nevertheless, these two objects are obviously different. From the distribution of the evaluations, it is easy to get that the object 1 is better than the object 2 because its intuitionistic decision potentials are more centralized, which is considered to be more reliable.

To depict the distribution (also called quality) of the intuitionistic decision potentials, Ren et al. [10] furtherly introduced the following definitions:

Definition 3.6 [10]. The quality of an intuitionistic decision potential given by decision-maker D_k can be measured by the similarity [16] between itself and the mean intuitionistic decision potential among all decision-makers:

$$Q_k = 1 - \frac{d(r_k, \overline{r})}{d(r_k, \overline{r}) + d(r_k^c, \overline{r})}, \quad k \in \{1, 2, \ldots, h\} \tag{3.14}$$

where $\overline{r}$ is the mean intuitionistic decision potential calculated by the IFWA operator. Especially, if the intuitionistic decision potentials given by all decision-makers are the same, then $Q_k = 1$ for $k = 1, 2, \ldots, h$, which indicates the highest compatibility among all decision-makers' opinions.

Definition 3.7 [10]. Intuitionistic decision exergy is the maximal effectiveness of decision potentials, expressed by:

$$B = Q\ E \tag{3.15}$$

Ren et al. [10] explained that the intuitionistic decision exergy is an indicator to measure the quality of the intuitionistic decision potentials. Considering that the common use of the entropy to depict the quality from the instability of a system, the intuitionistic decision entropy was introduced.

Definition 3.8 [10]. Intuitionistic decision entropy is to measure the unevenness of intuitionistic decision potentials of an alternative, which is given as:

$$S = E \ominus B \tag{3.16}$$

Based on the above concept, we present Fig. 3.1 to manifest the movement of an intuitionistic decision-making system.

If the intuitionistic decision potentials of an alternative provided by decision-makers are distributed more centralized, then the intuitionistic decision entropy of the alternative is smaller. Ren et al. [10] mentioned that intuitionistic decision entropy provides a feasible tool to assess the alternatives on both sides of quantity and quality, which makes the decision-making results more reliable and credible.

Based on the above work, the thermodynamic method for solving the MCDM problem in an intuitionistic fuzzy environment can be concluded [10]:

Step 1. Identify the alternatives A_i $(i = 1, 2, \ldots, m)$ and the criteria C_j $(j = 1, 2, \ldots, n)$ of the MCDM problem, and invite decision-makers D_k $(k = 1, 2, \ldots, h)$ to assess alternatives by providing the IFDMs:

$$R^k = (r_{ij}^k)_{m \times n}$$

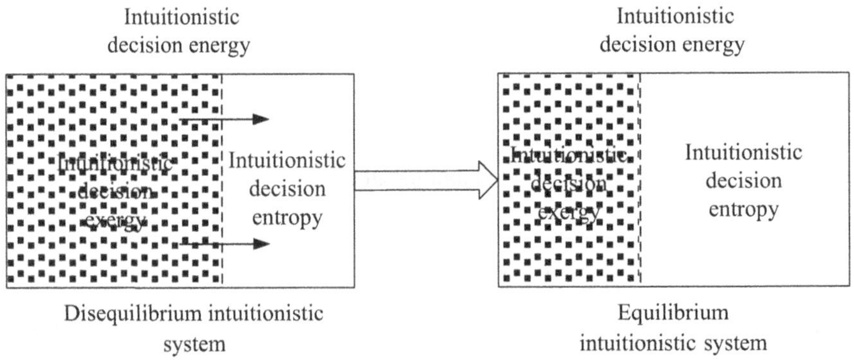

Fig. 3.1 Movement of an intuitionistic decision-making system

$$= \begin{pmatrix} (\mu_{11}^k, \nu_{11}^k) & (\mu_{12}^k, \nu_{12}^k) & \cdots & (\mu_{1n}^k, \nu_{1n}^k) \\ (\mu_{21}^k, \nu_{21}^k) & (\mu_{22}^k, \nu_{22}^k) & \cdots & (\mu_{2n}^k, \nu_{2n}^k) \\ \vdots & \vdots & \vdots & \vdots \\ (\mu_{m1}^k, \nu_{m1}^k) & (\mu_{m2}^k, \nu_{m2}^k) & \cdots & (\mu_{mn}^k, \nu_{mn}^k) \end{pmatrix} \quad \text{for} \quad k = 1, 2, \ldots, h \tag{3.17}$$

Step 2. Ask the decision-makers D_k $(k = 1, 2, \ldots, h)$ to determine the weights of the criteria w_j^k $(j = 1, 2, \ldots, n,\ k = 1, 2, \ldots, h)$, and calculate the intuitionistic decision energy matrix of each decision-maker:

$$E^k = \begin{pmatrix} w_1^k r_{11}^k & w_2^k r_{12}^k & \cdots & w_n^k r_{1n}^k \\ w_1^k r_{21}^k & w_2^k r_{22}^k & \cdots & w_n^k r_{2n}^k \\ \vdots & \vdots & \vdots & \vdots \\ w_1^k r_{m1}^k & w_2^k r_{m2}^k & \cdots & w_n^k r_{mn}^k \end{pmatrix} \quad \text{for} \quad k = 1, 2, \ldots, h \tag{3.18}$$

Step 3. Establish the quality matrix of each decision-maker:

$$Q^k = \begin{pmatrix} 1-\frac{d(r_{11}^k,\bar{r}_1^k)}{d(r_{11}^k,\bar{r}_1^k)+d((r_{11}^k)^c,\bar{r}_1^k)} & 1-\frac{d(r_{21}^k,\bar{r}_2^k)}{d(r_{21}^k,\bar{r}_2^k)+d((r_{21}^k)^c,\bar{r}_2^k)} & \cdots & 1-\frac{d(r_{n1}^k,\bar{r}_n^k)}{d(r_{n1}^k,\bar{r}_n^k)+d((r_{n1}^k)^c,\bar{r}_n^k)} \\ 1-\frac{d(r_{12}^k,\bar{r}_1^k)}{d(r_{12}^k,\bar{r}_1^k)+d((r_{12}^k)^c,\bar{r}_1^k)} & 1-\frac{d(r_{22}^k,\bar{r}_2^k)}{d(r_{22}^k,\bar{r}_2^k)+d((r_{22}^k)^c,\bar{r}_2^k)} & \cdots & 1-\frac{d(r_{n2}^k,\bar{r}_n^k)}{d(r_{n2}^k,\bar{r}_n^k)+d((r_{n2}^k)^c,\bar{r}_n^k)} \\ \vdots & \vdots & \vdots & \vdots \\ 1-\frac{d(r_{1m}^k,\bar{r}_1^k)}{d(r_{1m}^k,\bar{r}_1^k)+d((r_{1m}^k)^c,\bar{r}_1^k)} & 1-\frac{d(r_{2m}^k,\bar{r}_2^k)}{d(r_{2m}^k,\bar{r}_2^k)+d((r_{2m}^k)^c,\bar{r}_2^k)} & \cdots & 1-\frac{d(r_{mn}^k,\bar{r}_n^k)}{d(r_{mn}^k,\bar{r}_n^k)+d((r_{mn}^k)^c,\bar{r}_n^k)} \end{pmatrix} \tag{3.19}$$

$$\text{for} \quad k = 1, 2, \ldots, h$$

where $\bar{r}_j^k = \left(1 - \prod_{i=1}^m (1 - \mu_{ij}^k)^{\frac{1}{m}}, \prod_{i=1}^m (\nu_{ij}^k)^{\frac{1}{m}}\right)$.

Step 4. Construct the intuitionistic decision exergy matrices $B^k = \left[Q_{ij}^k\ E_{ij}^k\right]_{m \times n}$ of all decision-makers for $k = 1, 2, \ldots, h$.

Step 5. Obtain the averaging intuitionistic decision energy and the averaging intuitionistic decision exergy of the i - th alternative regarding the k - th decision-maker:

$$\overline{E}_i^k = \frac{1}{n}(E_{i1}^k \oplus E_{i2}^k \oplus \cdots \oplus E_{in}^k) \tag{3.20}$$

$$\overline{B}_i^k = \frac{1}{n}(B_{i1}^k \oplus B_{i2}^k \oplus \cdots \oplus B_{in}^k) \tag{3.21}$$

Step 6. Calculate the intuitionistic decision energy and the intuitionistic decision exergy of each alternative by

$$E_i = \frac{1}{k}(\overline{E}_i^1 \oplus \overline{E}_i^2 \oplus \cdots \oplus \overline{E}_i^l) \tag{3.22}$$

$$B_i = \frac{1}{k}(\overline{B}_i^1 \oplus \overline{B}_i^2 \oplus \cdots \oplus \overline{B}_i^l) \tag{3.23}$$

Step 7. Determine the intuitionistic decision entropy of each alternative:

$$S_i = E_i \ominus B_i \tag{3.24}$$

Step 8. Obtain the scores of each intuitionistic decision entropy $s(S_i)$ according to its score function, and rank alternatives. The smaller score of the intuitionistic decision entropy, the better the alternative.

Ren et al. [10] noted that the above thermodynamic method organically combines the quantity and the quality of the intuitionistic fuzzy information, which provides a rational and reliable alternative ranking.

3.4 Comparative Analyses

This section analyzes the differences among the results considering the distribution of intuitionistic information or not more intuitively.

3.4.1 Result Comparisons with Different Methods

Ren et al. [10] made comparisons among the method with the weighted aggregation operator [15], the TOPSIS [3], and the proposed thermodynamic method under an intuitionistic environment. For convenience, we denote the three methods as IFWAO, IF-TOPSIS, and IFTM, and the processes of IFWAO and IF-TOPSIS are presented at first:

Part 1. Decision-making process with IFWAO [15].

Step 1. Use IFWAO to aggregate the IFDMs provided by each decision-maker into an overall IFDM according to decision-makers' weights.

Step 2. Use IFWAO to aggregate the elements in the overall IFDM to obtain the final scores of alternatives according to the weights of criteria.

Step 3. Rank alternatives based on the final scores.

Part 2. Decision-making process with IF-TOPSIS [3].

Step 1. Use IFWAO to aggregate the IFDMs provided by each decision-maker into an overall IFDM according to decision-makers' weights.

Step 2. Identify the intuitionistic fuzzy positive ideal solution (PIS) and the intuitionistic fuzzy negative ideal solution (NIS) of the overall IFDM by

$$\begin{aligned} PIS &= \left\{C_j, \max_i \langle s(C_j(x_i))\rangle | j = 1, 2, \ldots, n\right\} \\ &= \{\langle C_1, (\mu_1^+, \nu_1^+)\rangle, \langle C_2, (\mu_2^+, \nu_2^+)\rangle, \ldots, \langle C_j, (\mu_n^+, \nu_n^+)\rangle\} \end{aligned} \tag{3.25}$$

and

$$\begin{aligned} NIS &= \left\{C_j, \min_i \langle s(C_j(x_i))\rangle | j = 1, 2, \ldots, n\right\} \\ &= \{\langle C_1, (\mu_1^-, \nu_1^-)\rangle, \langle C_2, (\mu_2^-, \nu_2^-)\rangle, \ldots, \langle C_j, (\mu_n^-, \nu_n^-)\rangle\} \end{aligned} \tag{3.26}$$

Step 3. Calculate the distance between the alternative A_i and the intuitionistic fuzzy PIS in the overall IFDM by $d(A_i, PIS) = \sum_{j=1}^{n} w_j d(C_j(A_i), C_j(PIS))$, and the distance between the alternative A_i and the intuitionistic fuzzy NIS by $d(A_i, NIS) = \sum_{j=1}^{j} w_j d(C_j(A_i), C_j(NIS))$, where w_j is the weight of the jth criterion.

Step 4. Ran alternatives according to the RC_i for $i = 1, 2, \ldots, m$, where

$$RC_i = \frac{d(A_i, NIS)}{d(A_i, PIS) + d(A_i, NIS)} \tag{3.27}$$

Ren et al. [10] proposed randomly generating the MCDM problems with IFNs for 1000 times in the comparative analyses and handling the problems generated by the methods with IFWAO, IF-TOPSIS, and IFTM, respectively. Without loss of generality, the generated MCDM problems contain five criteria, five alternatives, and five decision-makers. In the simulation process, the optimal selection with each decision-making method for each MCDM problem is recorded, and the decision-making results among the three methods can be compared in Fig. 3.2.

In Fig. 3.2, the items represented by each color have been marked, and some visualized conclusions can be obtained [10]: (1) among the 1000 simulations, the percentage of the optimal selections, derived from the three methods which are different, is 19%, but yet the ratio of reaching the same optimal selection is 7.3%, (2) according to the comparisons of the optimal results between any two methods, the ratios of the same selections obtained by the IFTM and the IFWAO, the IFTM and the IF-TOPSIS, the IFWAO and the IF-TOPSIS are 2%, 3.7%, and 68%, respectively. To this end, Ren et al. [10] summarized the below comparative conclusions:

(1) The ratio of obtaining the same optimal selections with the three decision-making methods is several times greater than the ratio of reaching the same results with the IFTM and the IFWAO (or the IF-TOPSIS). That is to say, if the optimal selections derived from the IFTM are the same as the selection derived from the method with the IFWAO (or the IF-TOPSIS), then it is probable that the three-decision making methods would derive the same optimal selections.

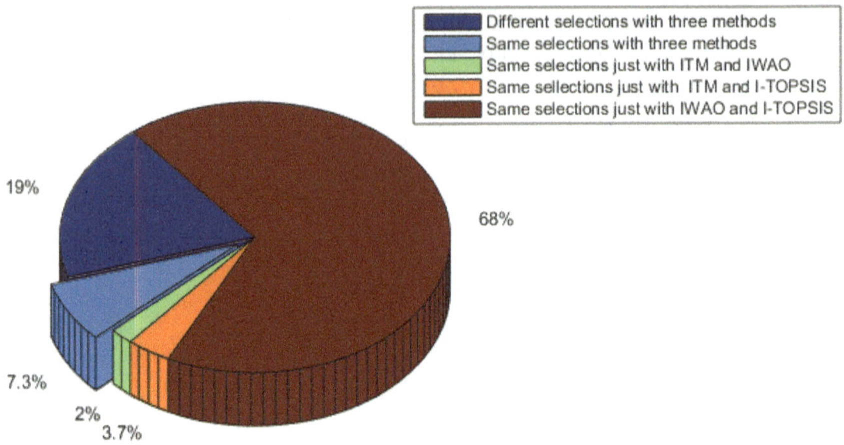

Fig. 3.2 The comparative results derived by three decision-making methods [10]

(2) The optimal selections with the IFTM are the most different from the optimal selections obtained by the IFWAO and the IF-TOPSIS, whereas the optimal selections derived from the method with the IFWAO and the IF-TOPSIS are the same in most cases. The essential difference between the IFTM and the other two methods is that the IFTM considers another important aspect: the distribution/quality of the intuitionistic fuzzy information. These simulation results show that taking the quality of the decision information into account often changes the decision-making results. In other words, the quality of decision information is a very significant character that we cannot ignore.

3.4.2 Sensitive Analysis on Results with Different Methods

To further investigate whether the simulation results in Fig. 3.2 can be influenced by the number of criteria or the number of alternatives, Ren et al. [10] designed the following simulation tests:

Let the number of criteria vary from 3 to 10, randomly generate the MCDM problems of five criteria with IFNs for 1000 times, and utilize the IFTM, the IFWAO, and the IF-TOPSIS to derive the optimal alternatives, respectively. We record the number of times when the optimal selections with any two decision-making methods are the same. Moreover, let the number of alternatives vary from 3 to 10 randomly conduct other 1000 times simulations for the MCDM problems of five criteria with IFNs and dispose of these MCDM problems like before. The simulation results can be shown in Figs. 3.3, 3.4 and 3.5.

By the above figures, some conclusions can be summed up [10]:

(1) Increasing the number of criteria hardly affects any two decision-making methods to obtain the same optimal selections.

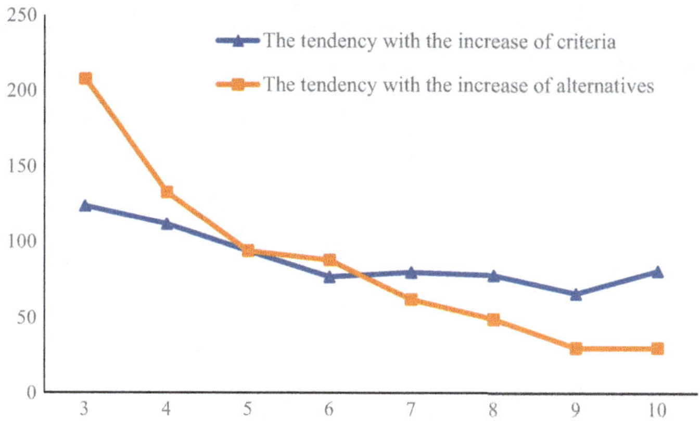

Fig. 3.3 Same selections with the IFTM and the IFWAO [10]

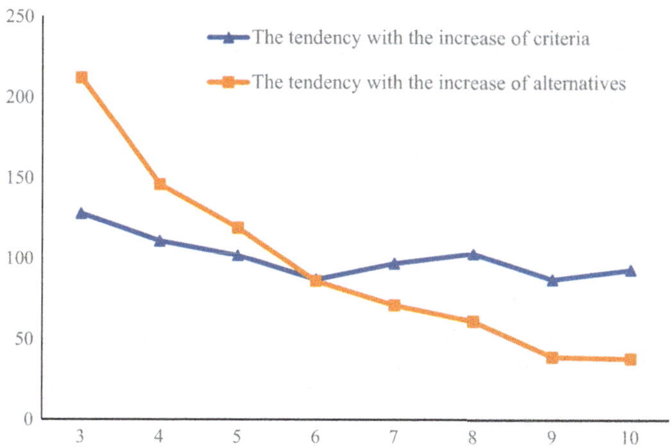

Fig. 3.4 Same selections with the IFTM and the IF-TOPSIS [10]

(2) The times for getting the same optimal selections with any two decision-making methods are significantly affected by the number of alternatives. With the increasing number of alternatives, getting the same optimal selections with the IFTM and the IFWAO (or the IF-TOPSIS) decreases, where this decreasing tendency between the method with the IFWAO and the IF-TOPSIS is unstable. The results indicate that the IFTM is more sensitive to the number of alternatives than the other two methods.

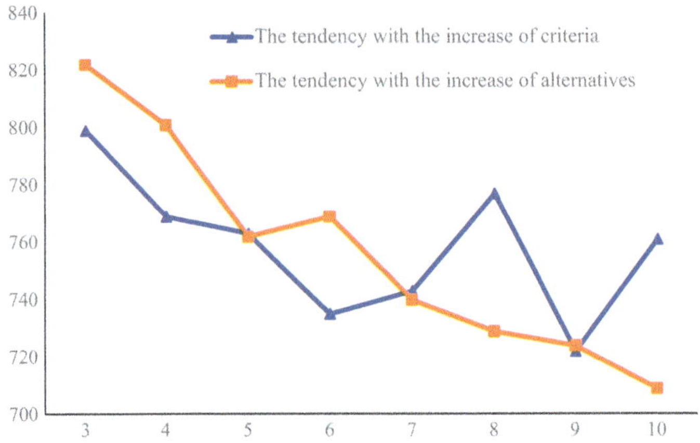

Fig. 3.5 Same selections with the method with the IFWAO and the IF-TOPSIS [10]

3.5 A Case Study on Addressing Hierarchical Diagnosis and Treatments

3.5.1 Description of the Case

To demonstrate the implementation process of IFTM, Ren et al. [10] simulated the application of supporting the hierarchical medical system of the West China Hospital and introduced the background as:

In the past few years, the problems in the Chinese medical industry, such as the uneven distribution of medical resource, the low efficiency of the medical industry, the unsound health-care system, cause the difficulty and the high cost of patients' treatments and such an issue is becoming more and more serious.

Recently, with the economic globalization of the twenty-first century, people's living standards rise ceaselessly. Meanwhile, the conflict between environmental issues and human development is also sharpening. Since 2013, extreme weather conditions, such as haze and typhoon, have emerged among the cities across China. Air quality indexes in these cities grossly exceed the standard. People's health has also been adversely affected. The research results of the global burden of diseases published in The Lancet indicate: Among the risk factors which affect the burden of diseases, outdoor air particulate pollution has reached as high as fourth in China, and about 20% causative factors in lung cancer are related to air pollution. Due to the increasingly prominent environmental issues, the operations of the medical industry in China is facing a severe test.

Nowadays, the number of sick with lung disease is increasing with the appalling rapidity of the weather conditions. Citizens are much more willing to seek medical advice and treatments in big hospitals due to their medical facilities and care conditions are much better than other hospitals. West China Hospital, which is one of the top hospitals in China, possesses advanced medical equipment and medical resource. The past couple of decades witnessed the increasing number of patients, which far excesses the capacity of West China Hospital to handle.

Under such a circumstance, the hierarchical medical system was introduced as an efficient way to release the pressure of the number of patients in West China Hospital, which aims to classify the difficulty of the treatment according to the disease of the light, heavy, slow, and urgent. Then, the different levels of medical institutions can undertake different kinds of diseases. Fundamentally, how to classify the different degrees of diseases is a critical problem in the hierarchical medical system.

Therefore, Ren et al. [10] focused on assisting the hierarchical medical system for lung diseases with the proposed method. The indicators associated with the diagnoses for each patient, who is possibly infected with lung diseases, were selected as:

- The signs of the body. The signs include the level of heart rate, the level of blood pressure, the level of glucose, and others. More un-normal signs are assigned a smaller value.
- The level of body temperature. Shivering and hyperthermia are two classic symptoms of pneumonia, which can be caused by a majority of lung diseases. More un-normal body temperature is assigned a smaller value.
- The frequency of cough. Lung diseases frequently cause bronchial damage, which is easy to produce cough. The symptoms like irritable cough, chronic cough, etc., are helpful to determine the elementary means of diagnosis. The higher the frequency of cough is assigned a smaller value.
- The frequency of hemoptysis. The cough may damage blood vessels on the surface layer of the lung and tumor. Different degrees of hemoptysis are a crucial indicator for judging the state of an illness. The higher the frequency of hemoptysis is assigned a smaller value.
- Extra-pulmonary manifestations of lung diseases. Some patients with early lung diseases may not manifest the symptoms of the lung diseases mentioned above, but they have other symptoms like arthralgia and arthrocele. More extra-pulmonary manifestations are assigned a smaller value.

According to the primary diagnosis, which considers these five criteria, the conditions and the critical degrees of patients can be judged. With these judgments, the patients are decided to be distributed to which levels and types of hospitals. As for the West China Hospital, the emergency patients should receive treatment in itself. The patients with less serious conditions should be treated in the second-tier hospitals. Other patients with light illness can seek medical advice and treatment at local hospitals.

In this case analysis, three doctors $D_k(k = 1, 2, 3)$ with the same weight were invited to diagnose diseases, the indicators were denoted as C_j $(j = 1, 2, 3, 4, 5)$, and the weight vector of the indicators was set as $w = (0.25, 0.15, 0.25, 0.25, 0.1)^T$. Suppose that there are five patients P_i $(i = 1, 2, 3, 4, 5)$ with lung diseases to be diagnosed and distributed, and the doctors gave their IFDMs as follows:

$$R^1 = \begin{pmatrix} (0.7, 0.2) & (0.7, 0.2) & (0.4, 0.3) & (0.6, 0.1) & (0.8, 0.1) \\ (0.5, 0.1) & (0.6, 0.2) & (0.5, 0.2) & (0.5, 0.1) & (0.9, 0.1) \\ (0.4, 0.4) & (0.4, 0.3) & (0.5, 0.3) & (0.4, 0.2) & (0.7, 0.1) \\ (0.8, 0.1) & (0.7, 0.2) & (0.6, 0.2) & (0.6, 0.2) & (0.9, 0.1) \\ (0.5, 0.4) & (0.5, 0.3) & (0.4, 0.4) & (0.4, 0.3) & (0.6, 0.2) \end{pmatrix},$$

$$R^2 = \begin{pmatrix} (0.7, 0.2) & (0.8, 0.2) & (0.5, 0.4) & (0.6, 0.3) & (0.7, 0.1) \\ (0.6, 0.1) & (0.6, 0.2) & (0.7, 0.2) & (0.6, 0.2) & (0.8, 0.1) \\ (0.5, 0.4) & (0.3, 0.2) & (0.6, 0.3) & (0.6, 0.3) & (0.7, 0.1) \\ (0.6, 0.1) & (0.7, 0.3) & (0.6, 0.2) & (0.6, 0.1) & (0.8, 0.1) \\ (0.5, 0.3) & (0.5, 0.3) & (0.5, 0.4) & (0.5, 0.4) & (0.7, 0.2) \end{pmatrix},$$

$$R^3 = \begin{pmatrix} (0.6, 0.1) & (0.6, 0.2) & (0.3, 0.3) & (0.6, 0.2) & (0.8, 0.1) \\ (0.7, 0.2) & (0.5, 0.3) & (0.7, 0.2) & (0.7, 0.3) & (0.7, 0.1) \\ (0.2, 0.3) & (0.6, 0.3) & (0.5, 0.2) & (0.6, 0.3) & (0.7, 0.1) \\ (0.6, 0.1) & (0.6, 0.1) & (0.7, 0.2) & (0.6, 0.1) & (0.8, 0.1) \\ (0.6, 0.4) & (0.6, 0.3) & (0.5, 0.4) & (0.4, 0.3) & (0.6, 0.3) \end{pmatrix}.$$

3.5.2 Decision-Making Process

Based on the doctors' IFDMs, the process of obtaining the overall results can be presented as follows [10]:

Step 1. Calculate the intuitionistic decision energy matrices of all doctors by Eq. (3.18):

$$E^1 = \begin{pmatrix} (0.2599,0.6687) & (0.1652,0.7855) & (0.1199,0.7401) & (0.2047,0.5623) & (0.1487,0.7943) \\ (0.1591,0.5623) & (0.1284,0.7855) & (0.1591,0.6687) & (0.1591,0.5623) & (0.2057,0.7943) \\ (0.1199,0.7953) & (0.0738,0.8348) & (0.1591,0.7401) & (0.1199,0.6687) & (0.1134,0.7943) \\ (0.3313,0.5623) & (0.1652,0.7855) & (0.2047,0.6687) & (0.2047,0.6687) & (0.2057,0.7943) \\ (0.1591,0.7953) & (0.0987,0.9348) & (0.1199,0.7953) & (0.1199,0.7401) & (0.0876,0.8513) \end{pmatrix}$$

$$E^2 = \begin{pmatrix} (0.2599,0.6687) & (0.2145,0.7855) & (0.1591,0.7953) & (0.2047,0.7401) & (0.1134,0.7943) \\ (0.2047,0.5623) & (0.1284,0.7855) & (0.2599,0.6687) & (0.2047,0.6687) & (0.1487,0.7943) \\ (0.1591,0.7953) & (0.0521,0.7855) & (0.2047,0.7401) & (0.2047,0.7401) & (0.1134,0.7943) \\ (0.2047,0.5623) & (0.1652,0.8348) & (0.2047,0.6687) & (0.2047,0.5623) & (0.1487,0.7943) \\ (0.1591,0.7401) & (0.0987,0.8348) & (0.1591,0.7953) & (0.1591,0.7953) & (0.1134,0.8513) \end{pmatrix}$$

$$E^3 = \begin{pmatrix} (0.2047,0.5623) & (0.1284,0.7855) & (0.0853,0.7401) & (0.2047,0.6687) & (0.1487,0.7943) \\ (0.2599,0.6687) & (0.0987,0.8348) & (0.2599,06687) & (0.2599,0.7401) & (0.1134,0.7943) \\ (0.0543,0.7401) & (0.1284,0.8348) & (0.1591,0.6687) & (0.2047,0.7401) & (0.1134,0.7943) \\ (0.2047,0.5623) & (0.1284,0.7079) & (0.2599,0.6687) & (0.2047,0.5623) & (0.1487,0.7943) \\ (0.2047,0.7953) & (0.1284,0.8348) & (0.1591,0.7953) & (0.1199,0.7401) & (0.0876,0.8866) \end{pmatrix}$$

Step 2. Obtain the mean intuitionistic decision energy of all doctors:

$$\overline{r}^1 = [(0.6319,0.1737),(0.5883,0.1320),(0.4651,0.2610),(0.7195,0.1569),(0.4644,0.3326)]^T$$

$$\overline{r}^2 = [(0.6553,0.2456),(0.6527,0.1569),(0.5531,0.2718),(0.6425,0.1402),(0.5249,0.3326)]^T$$

$$\overline{r}^3 = [(0.5707,0.1737),(0.6761,0.2195),(0.5113,0.2429),(0.6527,0.1189),(0.5319,0.3464)]^T$$

then establish the quality matrices Q^k $(k = 1, 2, 3)$:

$$Q^1 = \begin{pmatrix} 0.8478 & 0.8478 & 0.5587 & 0.8344 & 0.7884 \\ 0.8024 & 0.8545 & 0.8147 & 0.8024 & 0.7113 \\ 0.5000 & 0.7172 & 0.7639 & 0.6777 & 0.6515 \\ 0.8888 & 0.9265 & 0.8129 & 0.8129 & 0.8046 \\ 0.6190 & 0.8244 & 0.5000 & 0.6288 & 0.6634 \end{pmatrix}$$

$$Q^2 = \begin{pmatrix} 0.9090 & 0.7931 & 0.6218 & 0.8653 & 0.7923 \\ 0.8345 & 0.8957 & 0.8573 & 0.8957 & 0.8136 \\ 0.6403 & 0.5208 & 0.8137 & 0.8137 & 0.7251 \\ 0.8676 & 0.7204 & 0.8849 & 0.8676 & 0.8073 \\ 0.7963 & 0.7963 & 0.7130 & 0.7130 & 0.6772 \end{pmatrix}$$

$$Q^3 = \begin{pmatrix} 0.8647 & 0.8846 & 0.5000 & 0.8846 & 0.7321 \\ 0.9526 & 0.6811 & 0.9526 & 0.8215 & 0.8283 \\ 0.4495 & 0.7100 & 0.8518 & 0.7100 & 0.7078 \\ 0.8853 & 0.8853 & 0.8190 & 0.8853 & 0.8222 \\ 0.6758 & 0.7884 & 0.7413 & 0.5653 & 0.7884 \end{pmatrix}$$

Table 3.1 Decision-making results derived by the IFTM [10]

	Decision energy	Decision exergy	Decision entropy	Conditions of patients
Patient 1	(0.1764, 0.7211)	(0.1452, 0.7703)	(0.0365, 0.9360)	3
Patient 2	(0.1852, 0.6980)	(0.1592, 0.7389)	(0.0310, 0.9446)	1
Patient 3	(0.1334, 0.7628)	(0.0972, 0.8292)	(0.0401, 0.9199)	5
Patient 4	(0.2006, 0.6727)	(0.1730, 0.7123)	(0.0333, 0.9445)	2
Patient 5	(0.1322, 0.8046)	(0.0942, 0.8606)	(0.0420, 0.9349)	4

Step 3. Construct the intuitionistic exergy matrices $B^k = [Q_{ij}^k \;\, E_{ij}^k]_{m\times n}$ for $k = 1, 2, 3$:

$$B^1 = \begin{pmatrix} (0.2252,0.7110) & (0.1420,0.8149) & (0.0724,0.8376) & (0.1740,0.6186) & (0.1192,0.8340) \\ (0.1298,0.6301) & (0.1108,0.8136) & (0.1317,0.7205) & (0.1298,0.6301) & (0.1511,0.8489) \\ (0.0619,0.8918) & (0.0535,0.8785) & (0.1240,0.7946) & (0.0829,0.7614) & (0.0754,0.8607) \\ (0.3007,0.5995) & (0.1541,0.7996) & (0.1699,0.7210) & (0.1699,0.7210) & (0.1691,0.8309) \\ (0.1017,0.8678) & (0.0821,0.8617) & (0.0619,0.8918) & (0.0772,0.8276) & (0.0590,0.8987) \end{pmatrix}$$

$$B^2 = \begin{pmatrix} (0.2394,0.6937) & (0.1742,0.8258) & (0.1021,0.8673) & (0.1798,0.7707) & (0.0910,0.8332) \\ (0.1740,0.6185) & (0.1158,0.8055) & (0.2274,0.7083) & (0.1855,0.6974) & (0.1227,0.8392) \\ (0.1050,0.8636) & (0.0275,0.8819) & (0.1701,0.7828) & (0.1701,0.7828) & (0.0836,0.8462) \\ (0.1802,0.6069) & (0.1220,0.8780) & (0.1835,0.7004) & (0.1802,0.6069) & (0.1218,0.8304) \\ (0.1289,0.7869) & (0.0795,0.8661) & (0.1162,0.8493) & (0.1162,0.8493) & (0.0783,0.8967) \end{pmatrix}$$

$$B^3 = \begin{pmatrix} (0.1797,0.6079) & (0.1145,0.8077) & (0.0436,0.8603) & (0.1834,0.7005) & (0.1111,0.8449) \\ (0.2493,0.6816) & (0.0684,0.8843) & (0.2493,0.6816) & (0.2191,0.7809) & (0.0949,0.8264) \\ (0.0248,0.8735) & (0.0930,0.8796) & (0.1372,0.7098) & (0.1501,0.8076) & (0.0817,0.8496) \\ (0.1836,0.6007) & (0.1146,0.7366) & (0.2185,0.7193) & (0.1836,0.6007) & (0.1239,0.8275) \\ (0.1434,0.8566) & (0.1027,0.8673) & (0.1206,0.8438) & (0.0696,0.8435) & (0.0697,0.9094) \end{pmatrix}$$

Step 4. By Eqs. (3.20)–(3.24), decision-making results can be listed in Table 3.1.

Table 3.1 shows that the patient 3 is in the worst condition, who should be distributed to West China Hospital. Based on the availability of the accommodations for patients in each hospital, the other four patients can be arranged in different levels of hospitals [10].

3.5.3 *Comparison Case Results with Other Methods*

Ren et al. [10] further utilized the IFWAO [15] and IF-TOPSIS [3] to deal with the hierarchical medical system and compared the decision-making results (as shown in Table 3.2).

Table 3.2 The decision-making results with the IFTM, the IFWAO, and the IF-TOPSIS [10]

	IFTM		IFWAO		IF-TOPSIS	
	Final results	Conditions	Final results	Conditions	Final results	Conditions
Patient 1	(0.0365, 0.9360)	3	(0.9455, 0.0074)	3	0.70410	3
Patient 2	(0.0310, 0.9446)	1	(0.9537, 0.0045)	2	0.8955	2
Patient 3	(0.0401, 0.9199)	5	(0.8832, 0.0172)	4	0.3390	5
Patient 4	(0.0333, 0.9445)	2	(0.9652, 0.0026)	1	0.9678	1
Patient 5	(0.0420, 0.9349)	4	(0.8809, 0.0383)	5	0.4350	4

By Table 3.2, the determination of classifying patients is diversified with the three decision-making methods, in which the IFTM, the IFWAO, and the IF-TOPSIS indicate that Patient 3 and Patient 5 should be treated in West China Hospital. Furthermore, the rankings derived by the latter two methods are closer to each other, whereas the ranking derived by the proposed method is different from the other two rankings [10].

The IFTM obtains a ranking of the patients, which is majorly different from the rankings obtained by the method with the IFWAO and the IF-TOPSIS, the other two decision-making methods, which only make decisions according to the quantity of the data, present two more similar rankings of the patients. In most situations, the decision-makers' opinions are not the same or even different from each other. If only considering the values of the data rather than the distributions of the values, then we will overlook a significant character of the data and lose some useful decision information. From this point of view, the IFTM provides more credible results for decision-making problems [10].

3.6 Summary

With the increasing complexity of the practical problems, it is necessary to consider the uncertainty of the problem, excavate the potential value of the decision-making information, and establish a more reasonable decision-making process. As the knowledge structure of thermodynamics provides us with a tool to describe the stability (or instability) of a system, and intuitionistic fuzzy information provides us with a way to depict the fuzziness of objects, this chapter has introduced the thermodynamic method for MCDM with IFNs proposed by Ren et al. [10]. It obtains the results by combining the numerical value and distribution of intuitionistic fuzzy information in the decision-making process. The chapter has further presented experiments to show that the quality of IFSs affects the decision-making result, which makes contributions to the intuitionistic fuzzy decision theory.

References

1. Atanassov, K.T.: Intuitionistic fuzzy sets, In: Sgurev, V. (ed.) VII ITKR's Session (1983).
2. Atanassov, K.T.: Intuitionistic fuzzy set. Fuzzy Sets Syst. **20**, 87–96 (1986)
3. Boran, F.E., Genç, S., Kurt, M., Akay, D.: A multi-criteria intuitionistic fuzzy group decision making for supplier selection with TOPSIS method. Expert Syst. Appl. **36**, 11363–11368 (2009)
4. Bustince, H., Burillo, P.: Correlation of interval-valued intuitionistic fuzzy sets. Fuzzy Sets Syst. **74**(2), 237–244 (1995)
5. Chen, T.Y.: An interval-valued intuitionistic fuzzy LINMAP method with inclusion comparison possibilities and hybrid averaging operations for multiple criteria group decision making. Knowl.-Based Syst. **45**, 134–146 (2013)
6. Chen, T.Y.: The inclusion-based LINMAP method for multiple criteria decision analysis within an interval-valued Atanassov's intuitionistic fuzzy environment. Int. J. Inf. Technol. Decis. Mak. **13**, 1325–1360 (2014)
7. Devi, K.: Extension of VIKOR method in intuitionistic fuzzy environment for robot selection. Expert Syst. Appl. **38**, 14163–14168 (2011)
8. Lei, Q., Xu, Z.S.: Derivative and differential operations of intuitionistic fuzzy numbers. Int. J. Intell. Syst. **30**, 468–498 (2015)
9. Li, D.F.: Extension of the LINMAP for multi-attribute decision making under Atanassov's intuitionistic fuzzy environment. Fuzzy Optim. Decis. Making **7**, 17–34 (2008)
10. Ren, P.J., Xu, Z.S., Liao, H.C., Zeng, X.-J.: A thermodynamic method of intuitionistic fuzzy MCDM to assist the hierarchical medical system in China. Inf. Sci. **420**, 490–504 (2017)
11. Szmidt, E., Kacprzyk, J.: Distances between intuitionistic fuzzy sets. Fuzzy Sets Syst. **114**(3), 505–518 (2000)
12. Verma, M., Rajasankar, J.: A thermodynamical approach towards group multi-criteria decision making (GMCDM) and its application to human resource selection. Appl. Soft Comput. **52**, 323–332 (2017)
13. Wiecek, M.M., Ehrgott, M., Fadel, M., Figueira, J.R.: Multiple criteria decision making for engineering. Omega **36**, 337–339 (2008)
14. Wu, M.C., Chen, T.Y.: The ELECTRE multicriteria analysis approach based on Atanassov's intuitionistic fuzzy sets. Expert Syst. Appl. **38**, 12318–12327 (2011)
15. Xu, Z.S.: Intuitionistic fuzzy aggregation operations. IEEE Trans. Fuzzy Syst. **15**, 1179–1187 (2007)
16. Xu, Z.S., Chen, J.: An overview of distance and similarity measures of intuitionistic fuzzy sets. Int. J. Uncertain. Fuzziness Knowl.-Based Syst. **16**(4), 529–555 (2008)
17. Xu, Z.S., Liao, H.C.: Intuitionistic fuzzy analytic hierarchy process. IEEE Trans. Fuzzy Syst. **22**, 749–761 (2014)
18. Xu, Z.S., Yager, R.R.: Some geometric aggregation operators based on intuitionistic fuzzy sets. Int. J. Gen Syst **35**, 417–433 (2006)

19. Xu, Z.S., Yager, R.R.: Intuitionistic fuzzy Bonferroni means. IEEE Trans. Syst., Man, Cybern., Part B **41**(2), 568–578 (2011)
20. Zavadskas, E.K., Antucheviciene, J., Hajiagha, S.H.R., Hashemi, S.S.: Extension of weighted aggregated sum product assessment with interval-valued intuitionistic fuzzy numbers (WASPAS-IVIF). Appl. Soft Comput. **24**, 1013–1021 (2014)

Chapter 4
A Thermodynamic Method for Hesitant Fuzzy Decision Making Based on Prospect Theory

With the development of management and economic activities, it has been found that the expected utility theory [10, 17] sometimes does not conform to the management and economic behavior of human beings. Thus, Simon (1947) introduced the concept of bounded rationality to show the difficulty of achieving a perfect solution when people are limited by information and knowledge; instead, they prefer to find an optimal one among all existing alternatives. The emergence of "bounded rationality" guides that behavior decision making has stepped into a new development phase. Prospect theory [6], as a descriptive theory developed based on bounded rationality, can describe people's behavior when they are faced with risks. Later on, Gomes and Lima [5] proposed the TODIM method (an acronym in Portuguese for Interactive Multi-criteria Decision Making), which rank alternatives by calculating their partial superiorities and final superiorities according to the value function of prospect theory. Because the complexity of the decision making environment may cause the diversification of information expression, Wang and Sun [18] introduced the PROMETEE method based on prospect theory under trapezoidal fuzzy environment. Fan et al. [4] presented another method for MCDM problem based on prospect theory with real numbers and interval values.

As the generality of HFS defined by Torra [15] in describing the judgments among several possible values under uncertainty, its decision making theory and application have been largely discussed. In order to complete the decision making theory of HFS and expand the application value, it is necessary to combine the hesitant fuzzy information and prospect theory to develop methods. Furthermore, as described in Chap. 3, it is of great significance to fully mine the decision making information given by decision makers in an uncertain environment. The chapter will introduce a decision making method that utilizes the prospect theory to portray decision makers' psychological behaviors and applies the parameters in thermodynamics to take the numerical value and distribution of information into account.

P. Ren and Z. Xu, *Decision-Making Analyses with Thermodynamic Parameters and Hesitant Fuzzy Linguistic Preference Relations*, Studies in Fuzziness and Soft Computing 409, https://doi.org/10.1007/978-3-030-73253-0_4

4.1 Prospect Theory

Prospect theory [6] investigates the situation that people make a choice between alternatives under risk. It captures the characters, including reference dependence, diminishing sensitivity, and loss aversion of people, which manifest as the value function [6]:

$$v(x) = \begin{cases} x^{\alpha}, & x \geq 0 \\ -\lambda(-x)^{\beta}, & x < 0 \end{cases} \tag{4.1}$$

where x indicates the gains (when $x \geq 0$) or losses (when $x < 0$). $\alpha(0 \leq \alpha \leq 1)$ and $\beta(0 \leq \beta \leq 1)$ are two exponent parameters showing that the value function is concave for gains and convex for losses, respectively. $\lambda(\lambda \geq 1)$ means the risk aversion degree of people, the bigger value of λ represents that people are more risk-averse. Moreover, Tversky and Kahneman [16] have discussed that the outcomes are consistent with the original data when $\alpha = \beta = 0.88$ and $\lambda = 2.25$.

4.2 Hesitant Fuzzy Sets

A HFS on $X = \{x_1, x_2, \ldots, x_n\}$ is a function that when applied to returns a subset of [1] [15], and its mathematical symbol can be expressed as Xia and Xu [20]:

$$A = \{\langle x, h_A(x)\rangle | x \in X\} \tag{4.2}$$

where $h_A(x)$ is a set of several values in [1], which denotes the possible membership degrees of the element $x \in X$ to the set A. $h = h_A(x)$ is called a hesitant fuzzy element (HFE). Furthermore, the operations and aggregations of HFEs were investigated.

Definition 4.1 [20]. Let h, h_1 and h_2 be three HFEs, then

(1) $\lambda h = \bigcup_{\gamma \in h} \{1 - (1-\gamma)^{\lambda}\}$;
(2) $h_1 \oplus h_2 = \bigcup_{\gamma_1 \in h_1, \gamma_2 \in h_2} \{\gamma_1 + \gamma_2 - \gamma_1\gamma_2\}$;
(3) $h_1 \otimes h_2 = \bigcup_{\gamma_1 \in h_1, \gamma_2 \in h_2} \{\gamma_1\gamma_2\}$;
(4) $h_1 \ominus h_2 = \bigcup\limits_{\gamma_1 \in h_1, \gamma_2 \in h_2} \{\overline{\gamma}\}$, where $\overline{\gamma} = \begin{cases} \frac{\gamma_1 - \gamma_2}{1-\gamma_2}, & if\ \gamma_1 > \gamma_2\ and\ \gamma_2 \neq 1 \\ 0, & otherwise \end{cases}$.

Definition 4.2 [20]. Let $h_j (j = 1, 2, \ldots, n)$ be a collection of HFEs, and let HFWA: $\Theta^n \to \Theta$, then

$$HFWA(h_1, h_2, \ldots, h_n) = \mathop{\oplus}_{j=1}^{n} w_j h_j = \bigcup_{\gamma_1 \in h_1, \gamma_2 \in h_2, \ldots, \gamma_n \in h_n} \left\{1 - \prod_{j=1}^{n} (1-\gamma_j)^{w_j}\right\} \tag{4.3}$$

where $w = (w_1, w_2, \ldots, w_n)^T$ is the weight vector of $h_j (j = 1, 2, \ldots, n)$ with $0 \leq w_j \leq 1$ and $\sum_{j=1}^{n} w_j = 1$ for $j = 1, 2, \ldots, n$.

To compare HFEs, $s(h) = \frac{1}{\#h} \sum_{\chi=1}^{\#h} \gamma^{\chi}$ and $\overline{\sigma}(h) = \frac{1}{\#h} \left[\sum_{\chi=1}^{\#h} (\gamma^{\chi} - s(h))^2 \right]^{\frac{1}{2}}$ were introduced as the score function and the deviation degree of a HFE $h = \{\gamma^{\chi} | \chi = 1, 2, \ldots, \#h\}$ [9, 20]. The comparative results of two HFEs h_1 and h_2 can be obtained by the rules [1]:

(1) If $s(h_1) < s(h_2)$, then $h_1 < h_2$;
(2) If $s(h_1) = s(h_2)$, then

- If $\overline{\sigma}(h_1) < \overline{\sigma}(h_2)$, then $h_1 > h_2$;
- If $\overline{\sigma}(h_1) = \overline{\sigma}(h_2)$, then $h_1 = h_2$.

For better application of hesitant fuzzy information, Xu and Zhang [24] defined an extension method for a HFE $h = \{\gamma^{\chi} | \chi = 1, 2, \ldots, \#h\}$: suppose that the maximum and minimum elements in h are γ^+ and γ^-, respectively, then the extended elements can be got by $\gamma^* = \theta\gamma^+ + (1 - \theta)\gamma^-$ $(0 \leq \theta \leq 1)$.

Definition 4.3 [9]. Let $h_1 = \{\gamma_1^{\chi} | \chi = 1, 2, \ldots, \#h\}$ and $h_2 = \{\gamma_2^{\chi} | \chi = 1, 2, \ldots, \#h\}$ be two HFEs, then the hesitant normalized Hamming distance of them can be calculated by.

$$d(h_1, h_2) = \frac{1}{\#h} \sum_{\chi=1}^{\#h} \left| \gamma_1^{\chi} - \gamma_2^{\chi} \right| \tag{4.4}$$

4.3 Hesitant Fuzzy Prospect Matrix

This section aims to construct a hesitant fuzzy prospect matrix (HFPM) of a hesitant fuzzy decision matrix (HFDM), which includes applying the prospect theory to handle the original information in HFDM, and transferring the information into the judgments with the decision makers' attitudes towards gains and losses [12].

During the decision making process, different decision makers may have different expectations for alternatives. For the same judgment of an alternative, a decision maker may satisfy its performance, but it is probably that another doesn't satisfy. As prospect theory is proposed to portray the discrepant perception of decision makers towards the alternatives performing better or worse, in the following, we present concepts and transformation rules to address such a situation.

Definition 4.4 [12] The expectation-level of a decision maker is regarded as his/her reference point. If the judgment in HFDM is higher than the expectation-level, it is viewed as the gain; adversely, it should be taken for the loss.

Based on the concept of reference point, Ren et al. [12] put forward to apply the value function of prospect theory to obtain the HFPM, and called the outcome derived by comparing the judgment and the expectation-level as a hesitant fuzzy prospect judgment (HFPJ).

Let h be a HFE and $\tilde{h}$ be the corresponding expectation level, where is h a judgment in HFDM given by a decision maker, Ren et al. [12] discussed the following cases:

Case 1. $\gamma_i \geq \widetilde{\gamma}_i$ and $\widetilde{\gamma}_i \neq 1$, where $\gamma_i \in h$ and $\widetilde{\gamma}_i \in \tilde{h}$.

$\gamma_i \geq \widetilde{\gamma}_i$ indicates that the decision maker thinks that the performance of the alternative reaches his/her expectation. In such a case, the corresponding HFPJ can be calculated by $\frac{\gamma_i - \widetilde{\gamma}_i}{1-\widetilde{\gamma}_i}$.

Case 2. $\gamma_i = \widetilde{\gamma}_i = 1$, where $\gamma_i \in h_i$ and $\widetilde{\gamma}_i \in \tilde{h}$.

When $\gamma_i = \widetilde{\gamma}_i = 1$, the judgment of the alternative is equal to the decision maker's expectation, the corresponding HFPJ is 0.

Case 3. $\gamma_i < \widetilde{\gamma}_i$, where $\gamma_i \in h_i$ and $\widetilde{\gamma}_i \in \tilde{h}$.

When $\gamma_i < \widetilde{\gamma}_i$, the judgment should be seen as the loss. Considering that a negative number usually describes the loss, but there doesn't exist negative expression in the hesitant fuzzy environment, Ren et al. [12] gave the following process to depict the loss:

Stage 1. Transforming the judgment. Let $\tau = \frac{\widetilde{\gamma}_i - \gamma_i}{1-\gamma_i}$, then, the loss can be regarded as a monotone decreasing function of τ.
Stage 2. Determining the function. The function must satisfy the following conditions: (1) The function strictly decreases when τ increases; (2) The range of the function is located in [0, 1]. To meeting the requirement, the negative exponential function is adopted.
Stage 3. Calculating the loss. Based on the above stages, the following formula can be used to compute the loss

$$f(\gamma_i) = e^{-\left(\frac{\widetilde{\gamma}_i - \gamma_i}{1-\gamma_i}\right)} - e^{-1} \tag{4.5}$$

Ren et al. [12] noted that the range of Eq. (4.5) belongs to the interval [0, 1], and stated that it is reasonable to establish a function reflecting the relationship between the judgment and the loss with the diminishing marginal utility. They furtherly explained why using the function $f(\gamma_i) = e^{-\left(\frac{\widetilde{\gamma}_i - \gamma_i}{1-\gamma_i}\right)} - e^{-1}$ instead of the function $f(\gamma_i) = e^{-\left(\frac{\widetilde{\gamma}_i - \gamma_i}{1-\gamma_i}\right)}$ is that $e^{-\left(\frac{\widetilde{\gamma}_i - \gamma_i}{1-\gamma_i}\right)} \geq e^{-1} = 0.3679$, which far exceeds 0, which indicates $e^{-\left(\frac{\widetilde{\gamma}_i - \gamma_i}{1-\gamma_i}\right)}$ lacks rationality to represent the losses.

The above transformation techniques can be summarized as follows:

Rule 4.1 [12]. Let h be a HFE, and $\tilde{h}$ be the corresponding expecting HFE, then the HFPJ h_υ can be obtained by

(1) If $\gamma_i \geq \widetilde{\gamma}_i$ and $\widetilde{\gamma}_i \neq 1$, where $\gamma_i \in h$ and $\widetilde{\gamma}_i \in \tilde{h}$, then $\gamma_{\upsilon,i} = \left(\frac{\gamma_i - \widetilde{\gamma}_i}{1-\widetilde{\gamma}_i}\right)^\alpha$, $\gamma_{\upsilon,i} \in h_\upsilon$,

(2) If $\gamma_i = \widetilde{\gamma}_i = 1$, where $\gamma_i \in h_i$ and $\widetilde{\gamma}_i \in \tilde{h}$, then $\gamma_{\upsilon,i} = 0$, $\gamma_{\upsilon,i} \in h_\upsilon$,

(3) If $\gamma_i < \widetilde{\gamma}_i$, where $\gamma_i \in h_i$ and $\widetilde{\gamma}_i \in \tilde{h}$, then $\gamma_{\upsilon,i} = \frac{1}{\lambda}\left(e^{-\left(\frac{\widetilde{\gamma}_i - \gamma_i}{1-\gamma_i}\right)} - e^{-1}\right)^\beta$, $\gamma_{\upsilon,i} \in h_\upsilon$,

where $0 \leq \alpha, \beta \leq 1$ and $\lambda \geq 1$.

It should be noted that the HFPJs derived by Rule 4.1 construct a HFE [12].

4.4 Thermodynamic Decision-Making Method Based on Prospect Theory

In this section, we present the thermodynamic decision making method based on the HFPMs constructed in Sect. 4.3. Firstly, some basic definitions can be given:

Definition 4.5 [12]. Let the hesitant decision potential be the energy of an alternative with respect to a criterion, which is represented by a HFPJ, and let the hesitant decision force be the weight of the criterion.

Definition 4.6 [12]. The hesitant decision energy is an indicator to manifest the capacity an alternative possesses in the decision making process, which can be calculated by.

$$E = w\,h_\upsilon \tag{4.6}$$

where h_υ is the hesitant decision potential of an alternative and w is the corresponding hesitant decision force.

We draw a graph to show the relationship among the three above concepts: (Fig. 4.1)

Similar to the decision making with intuitionistic fuzzy information, we can use the hesitant decision potential, hesitant decision force, and hesitant decision energy to obtain the score of alternatives by the methods of aggregation, TOPSIS, etc. But they all ignore the distribution/quality of hesitant fuzzy information, which is a crucial property in the decision making process and improves the dependence of the results. In what follows, we introduce an efficient way to describe the structure of the decision making procedure with HFSs:

Definition 4.7 [12]. Let the quality of a hesitant decision potential be the similarity to the mean hesitant decision potential, and can be measured by Xu and Xia [23].

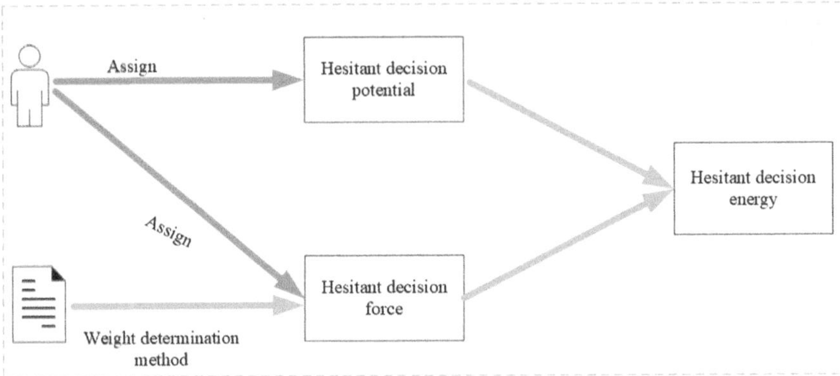

Fig. 4.1 The relationship among hesitant decision potential, force, and energy

$$q = 1 - \frac{1}{\#h}\sum_{\chi=1}^{\#h}\left|(h_{v,i})^{\chi} - (\overline{h}_{v,i})^{\chi}\right| \tag{4.7}$$

where $\overline{h}_{v,i}$ is the mean hesitant decision potential, which can be obtained by the HFWA operator. When the quality approaches 1, it possesses relatively good quality; adversely, it has relatively lousy quality.

The quality depicts the dependability of the hesitant decision potentials from the aspect of the internal structure, and the following concept is presented to describe the usefulness of the hesitant decision potential:

Definition 4.8 [12]. Let the hesitant decision exergy be the maximum effectiveness of the hesitant decision potential, which depicts its capacity of working the best, expressed by:

$$B = qE \tag{4.8}$$

The hesitant decision exergy describes the hesitant decision potential from the stable aspect; and for the unstable aspect, another definition can be given:

Definition 4.9 [12]. Let the hesitant decision entropy be an indicator to describe the unevenness of the hesitant decision potential of an alternative, which is the opposite of the hesitant decision exergy and represented by

$$S = E \ominus B \tag{4.9}$$

It is noted that the hesitant decision exergy and the hesitant decision entropy are both effective to be used to make the decision. So far, we can briefly present the steps of the thermodynamic decision making method based on prospect theory under a hesitant fuzzy environment [12]:

Step 1. Identify the alternatives $A_i (i = 1, 2, \ldots, m)$ and the criteria $C_j (j = 1, 2, \ldots, n)$, and determine the decision makers $D_k (k = 1, 2, \ldots, d)$ in the decision making problem.
Step 2. Invite each decision maker to assign the weights of criteria as w_j^k for $j = 1, 2, \ldots, n$ and $k = 1, 2, \ldots, d$, where w_j^k indicates the weight of the criterion C_j assigned by the decision maker D_k.
Step 3. Invite each decision maker to evaluate alternatives with respect to all criteria, and give their HFDMs:

$$H^k = (h_{ij}^k)_{m\times n} = \begin{pmatrix} h_{11}^k & h_{12}^k & \cdots & h_{1n}^k \\ h_{21}^k & h_{22}^k & \cdots & h_{2n}^k \\ \vdots & \vdots & \vdots & \vdots \\ h_{m1}^k & h_{m2}^k & \cdots & h_{mn}^k \end{pmatrix} \quad \text{for } k = 1, 2, \ldots, d \tag{4.10}$$

Step 4. Construct the HFPMs $H_v^k = (h_{v,ij}^k)_{m\times n}$ for $k = 1, 2, \ldots, d$ by Rule 4.1.
Step 5. Obtain the hesitant decision energy matrices and the quality matrices of decision makers by Eq. (2.6) and Eq. (2.7), respectively, which are

$$E^k = \begin{pmatrix} w_1^k h_{v,11}^k & w_2^k h_{v,12}^k & \cdots & w_n^k h_{v,1n}^k \\ w_1^k h_{v,21}^k & w_2^k h_{v,22}^k & \cdots & w_n^k h_{v,2n}^k \\ \vdots & \vdots & \vdots & \vdots \\ w_1^k h_{v,m1}^k & w_2^k h_{v,m2}^k & \cdots & w_n^k h_{v,mn}^k \end{pmatrix} \quad \text{for } k = 1, 2, \ldots, d \tag{4.11}$$

and

$$q^k = \begin{pmatrix} 1-\frac{1}{\#h}\sum_{\chi=1}^{\#h}\left|(h_{v,11}^k)^\chi-(\bar{h}_{v,1})^\chi\right| & 1-\frac{1}{\#h}\sum_{\chi=1}^{\#h}\left|(h_{v,12}^k)^\chi-(\bar{h}_{v,1})^\chi\right| & \cdots & 1-\frac{1}{\#h}\sum_{\chi=1}^{\#h}\left|(h_{v,1n}^k)^\chi-(\bar{h}_{v,1})^\chi\right| \\ 1-\frac{1}{\#h}\sum_{\chi=1}^{\#h}\left|(h_{v,21}^k)^\chi-(\bar{h}_{v,2})^\chi\right| & 1-\frac{1}{\#h}\sum_{\chi=1}^{\#h}\left|(h_{v,22}^k)^\chi-(\bar{h}_{v,2})^\chi\right| & \cdots & 1-\frac{1}{\#h}\sum_{\chi=1}^{\#h}\left|(h_{v,2n}^k)^\chi-(\bar{h}_{v,2})^\chi\right| \\ \vdots & \vdots & \vdots & \vdots \\ 1-\frac{1}{\#h}\sum_{\chi=1}^{\#h}\left|(h_{v,m1}^k)^\chi-(\bar{h}_{v,m})^\chi\right| & 1-\frac{1}{\#h}\sum_{\chi=1}^{\#h}\left|(h_{v,m2}^k)^\chi-(\bar{h}_{v,m})^\chi\right| & \cdots & 1-\frac{1}{\#h}\sum_{\chi=1}^{\#h}\left|(h_{v,mn}^k)^\chi-(\bar{h}_{v,m})^\chi\right| \end{pmatrix}$$

$$\text{for } k = 1, 2, \ldots, d \tag{4.12}$$

where $\bar{h}_{v,1} = \bigcup_{\gamma_{v,i1}\in h_{v,i1}, \gamma_{v,i2}\in h_{v,i2}, \ldots, \gamma_{v,in}\in h_{v,in}} \left\{1 - \prod_{j=1} (1 - \gamma_{v,ij})^{w_j}\right\}$.

Step 6. Calculate the hesitant decision exergy matrices for each decision maker:

$$B^k = \left[q_{ij}^k \ E_{ij}^k\right]_{m\times n} \quad \text{for } k = 1, 2, \ldots, d \tag{4.13}$$

Step 7. Acquire the averaging hesitant decision energy and the averaging hesitant decision exergy of the alternative A_i as regards the decision maker D_k:

$$\overline{E}_i^k = \frac{1}{n}(E_{i1}^k \oplus E_{i2}^k \oplus \cdots \oplus E_{in}^k) \text{ for } i = 1, 2, \ldots, m \text{ and } k = 1, 2, \ldots, d \tag{4.14}$$

$$\overline{B}_i^k = \frac{1}{n}(B_{i1}^k \oplus B_{i2}^k \oplus \cdots \oplus B_{in}^k) \text{ for } i = 1, 2, \ldots, m \text{ and } k = 1, 2, \ldots, d \tag{4.15}$$

Step 8. Determine the hesitant decision entropy of each alternative by

$$S_i = E_i \ominus B_i \text{ for } i = 1, 2, \ldots, m \tag{4.16}$$

where E_i and B_i are the overall hesitant decision energy and the overall hesitant decision exergy of the alternative A_i, respectively, which are derived by

$$E_i = \frac{1}{d}(\overline{E}_i^1 \oplus \overline{E}_i^2 \oplus \cdots \oplus \overline{E}_i^d) \text{ for } i = 1, 2, \ldots, m \tag{4.17}$$

$$B_i = \frac{1}{d}(\overline{B}_i^1 \oplus \overline{B}_i^2 \oplus \cdots \oplus \overline{B}_i^d) \text{ for } i = 1, 2, \ldots, m \tag{4.18}$$

go to the next step;

Step 9. Rank alternatives, the smaller value of S_i, the better the alternative.

Ren et al. [12] presented Fig. 4.2 to describe the above decision making procedure.

4.5 Discussions

4.5.1 The Validation of the Method

The analysis of method validity helps to better mine its applicability and has significance on its practical application. Motivated by the experiment given by Chiclana et al. [2], Ren et al. [12] designed a procedure for verifying the validity of the proposed method.

- *Experimental content:* Compare the results of multiple criteria decision making (MCDM) problems derived by hesitant fuzzy TOPSIS [24], hesitant fuzzy TODIM [25], and the proposed method, and summarize conclusions based on the comparisons.

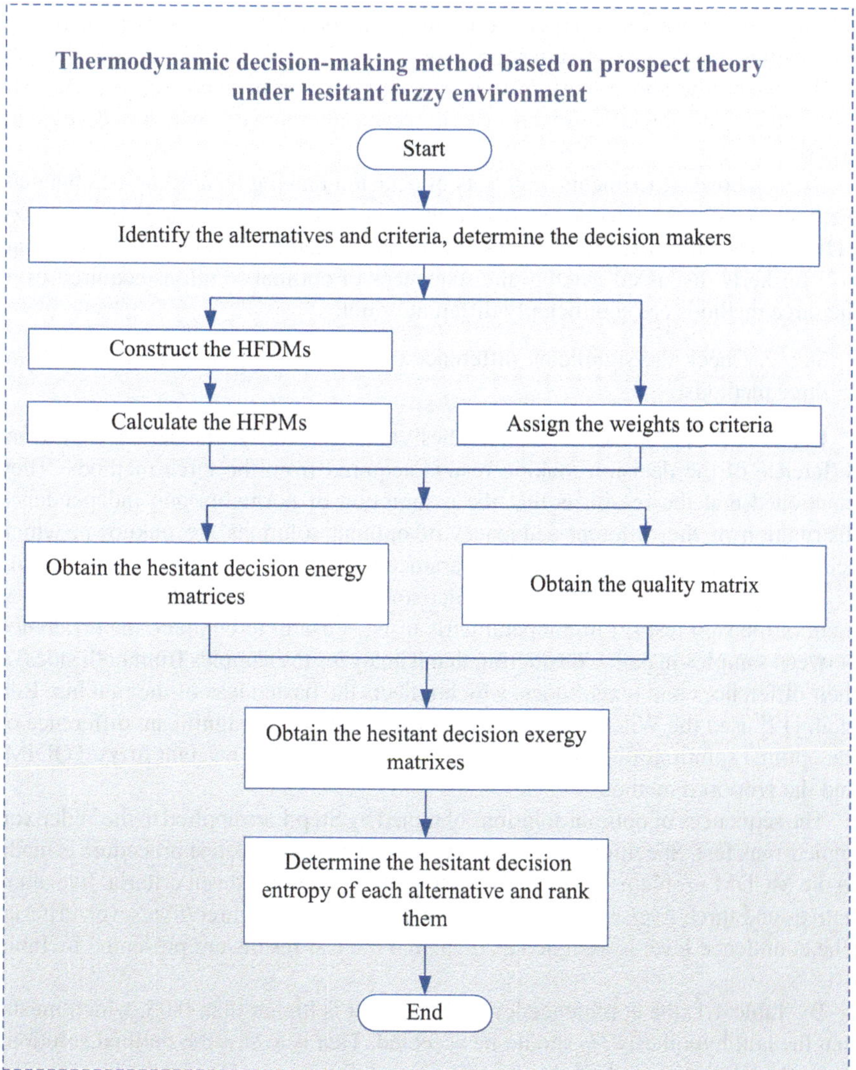

Fig. 4.2 Procedure of the proposed method [12]

- *Experimental design:*

  Step 1. Obtain the decision making results of hesitant fuzzy TOPSIS, hesitant fuzzy TODIM, and the proposed method, respectively.

We randomly produce 1000 MCDM problems with HFSs, containing three alternatives and three criteria. Then we apply hesitant fuzzy TOPSIS, hesitant fuzzy TODIM, and the proposed method to solve the 1000 problems, and record the optimal solutions of the problems. Three sequences of the optimal solutions with respect to each method can be obtained, respectively.

We extend the above work to the decision scenarios of five/seven criteria, five alternatives, and three/five/seven criteria, seven alternatives, and three/five/seven criteria.

By the above experiments, nine sets of decision making results can be obtained. Each set contains three sequences of optimal solutions derived by using hesitant fuzzy TOPSIS, hesitant fuzzy TODIM, and the proposed method. Based on this, Ren et al. [12] furtherly discussed whether the sequences of optimal solutions acquired from the three methods are significantly different or not.

Step 2. Check the significant difference of the optimal solutions acquired from three methods.

Ren et al. [12] proposed to use hypothesis testing to verify whether the significant difference of the decision making results acquired from the three methods. They mentioned that the reason is that the assumption of normality and independence distribution of the different sequences of optimal solutions are unknown, which accords with the condition of the nonparametric test.

For any two related samples, the Wilcoxon signed-rank test [19], as an improvement of the sign test [7] in nonparametric tests, is useful to compare the difference between samples in pair. Considering that it analyzes the samples from both sides of their differences and magnitudes, which reflects the basic ideas of the ranking, Ren et al. [12] used the Wilcoxon signed-rank test to justify the significant difference of the optimal solutions obtained from hesitant fuzzy TOPSIS, hesitant fuzzy TODIM, and the proposed method.

The sequences of optimal solutions obtained by Step 1 are applied to the Wilcoxon signed-rank test. Specifically, the same Wilcoxon signed-rank test procedure is made to the MCDM problems of three alternatives and three/five/seven criteria, five alternatives and three/five/seven criteria, seven alternatives and three/five/seven criteria. The confidence level is assigned as 0.95, and the test results are presented in Table 4.1.

By Table 4.1, the confidence level of each test is higher than 0.05, which means that the null hypothesis H_0 should be accepted. That is to say, the optimal solutions derived by different methods have no significant differences. However, observing the optimal sequences obtained by the experiments. It is easy to find that the different methods usually lead to different decision making results for each MCDM problem [12]. Furthermore, a figure of manifesting the experiment data can be shown in Fig. 4.3.

Figure 4.3 records the ratio of same optimal selections of the 1000 MCDM problems with five alternatives and five criteria, Ren et al. [12] noted that for each number of alternatives and each number of criteria in the experiments, the ratios of the same rankings obtained by any two methods are closed to the ratios in Fig. 4.3. Some comparative conclusions can be drawn [12]:

Table 4.1 The Wilcoxon signed-rank test results of the optimal solutions obtained by hesitant fuzzy TOPSIS, hesitant fuzzy TODIM, and the proposed method [12]

	hesitant fuzzy TODIM	The proposed method
3A3C Hesitant fuzzy TOPSIS Hesitant fuzzy TODIM	Sig. 0.888	Sig. 0.880 Sig. 0.718
3A5C Hesitant fuzzy TOPSIS Hesitant fuzzy TODIM	Sig. 0.921	Sig. 1.000 Sig. 0.944
3A7C Hesitant fuzzy TOPSIS Hesitant fuzzy TODIM	Sig. 0.995	Sig. 0.974 Sig. 0.946
5A3C Hesitant fuzzy TOPSIS Hesitant fuzzy TODIM	Sig. 0.902	Sig. 0.973 Sig. 0.980
5A5C Hesitant fuzzy TOPSIS Hesitant fuzzy TODIM	Sig. 0.914	Sig. 0.998 Sig. 0.938
5A7C Hesitant fuzzy TOPSIS Hesitant fuzzy TODIM	Sig. 0.942	Sig. 0.875 Sig. 0.844
7A3C Hesitant fuzzy TOPSIS Hesitant fuzzy TODIM	Sig. 0.907	Sig. 0.923 Sig. 0.961
7A5C Hesitant fuzzy TOPSIS Hesitant fuzzy TODIM	Sig. 0.998	Sig. 0.982 Sig. 0.965
7A7C Hesitant fuzzy TOPSIS Hesitant fuzzy TODIM	Sig. 0.945	Sig. 0.942 Sig. 0.915

Note mAnC represents the MCDM problems of m alternatives and n criteria

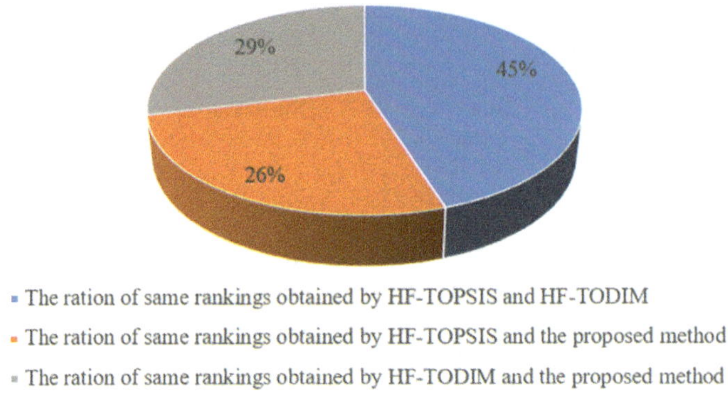

Fig. 4.3 The ratio of same optimal selections obtained by any two methods [12] (*Note* HF-TOPSIS represents the hesitant fuzzy TOPSIS, HF-TODIM represents the hesitant fuzzy TODIM)

(1) The Wilcoxon signed-rank test shows the sequences of the optimal selections obtained by the hesitant fuzzy TOPSIS, the hesitant fuzzy TODIM, and the proposed method have no significant differences. But the ratios of the same optimal selections obtained by any two methods are different. Specifically, the hesitant fuzzy TOPSIS and the hesitant fuzzy TODIM are more likely to obtain the same optimal selections, while the hesitant fuzzy TOPSIS and the proposed method, the hesitant fuzzy TODIM, and the proposed method are not.

(2) The time complexity relationship of the three methods is $o(HF-TODIM) >> o(The\ proposed\ method) > o(HF-TOPSIS)$, where "$>>$" indicates much bigger and "$>$" indicates a bit bigger.

(3) For a specific decision making problem, we probably get three different optimal solutions by the three different methods. Hence, it is imperative to know what methods you should choose. For most decision makers, they are willing to make decisions in a simple way. Hence, among the three decision making methods, the hesitant fuzzy TOPSIS and the proposed method are two great choices. To get a cursory solution, the hesitant fuzzy TODIM is useful. However, the proposed method is much more appropriate to acquire a deliberate ranking, especially in the problems containing much uncertainty and complexity.

4.5.2 Comments on the Existing Method

To clearly manifest the characteristics of the proposed method, Ren et al. [12] further made discussions among several existing methods and the proposed method.

There exist some researches on addressing the decision making problems based on people's habits. Saaty and Shang [14] proposed a method to sort and arrange the mental thoughts and criteria in the widely heterogeneous entities into homogeneous groups. Considering that the relative values are more instructive than a single value,

Saaty [13] utilized the concept of 'dependent' in AHP and ANP to rate alternatives with respect to an ideal alternative. Since the dynamic decision making problems often appear in practices, the methods for multi-period MCDM with techniques of determining dynamic weighted operators were introduced [21, 22]. Moreover, Pérez et al. [11] created a dynamic method to allow decision makers to remove old inferior alternatives and insert new superior alternatives at each period of the decision making process. Dong et al. [3] provided a method with the characteristics that decision makers can use their individual sets of criteria and dynamically change the individual sets of criteria and alternatives.

Ren et al. [12] showed that because the proposed method processes the ability to simulate the actual decision making problems and consider people's thoughts, it is meaningful to comment on the proposed method and the MCDM methods mentioned above. More specifically,

Q1. Does the method consider people's decision habits and thought?
Q2. Does the method simulate the actual decision making situation well?
Q3. Does the method increase its effectiveness and applicability by paying more attention to alternatives, criteria, or weights?

The conclusions can be addressed: (1) the methods in [21, 22] ignore Q1, (2) the methods in [13, 14] ignore Q2, (3) the methods in [3, 11], and the proposed method take all above aspects into accounts [12].

4.6 A Case Study on Emergency Decision Making in Firing and Exploding Accident

4.6.1 Description of the Case

Ren et al. [12] applied the proposed method to select the emergency response for the firing and exploding in Tianjin Binhai, and they briefly presented the background of the case as follows:

On August 12, 2015, the hazardous chemical substances of Ruihai logistics outbroke of fire at Port Group in Tianjin Binhai New Area, and the explosion happened in firefighting and rescue work. As of 3:00 p.m. on September 11, the death toll had risen to 165. Hardly can anyone deny that this accident caused heavy losses to urban resident life and property and significantly impacted life circumstances in the city and its surrounding areas. In this accident, the fire impaired thousands of vehicles, and the wave of the explosion gave rise to the damage of the residential quarters around the site; the cyanide in the sewage of the core area of the explosion averagely exceeded 40 times. Under such a situation, establishing an effective decision making method in the emergency response is crucial to avoid additional losses.

The three aspects are crucial for the emergency response: (1) Organize to rescue victims. When the accident occurs, the emergency rescue teams should first take action to rescue the injured and trapped people, then evacuate the people around the firing and exploding site. (2) Monitor the situation quickly. In the emergency rescue procedure, it is necessary to take adequate measures to monitor the situation and detect the accident. (3) Prevent other explosions. The explosion is of great powers to destroy surroundings, which can broadly impact people's life. Therefore, it is necessary to handle the explosive goods duly and prevent more explosions.

Ren et al. [12] supposed that in the emergency response of the firing and exploding, three rescue projects (denoted by $P_i (i = 1, 2, 3)$) are considered by two decision makers (denoted by $D^k (k = 1, 2)$). They should choose a rescue project, which probably performs best with comprehensive thinking over the above three critical aspects (denoted by C_j $(j = 1, 2, 3)$. Assume that the two decision makers have the same weights, they assign the weights for the considered aspects as $w = (0.4, 0.3, 0.3)^T$ and assign the expectation-level of each project with respect to each aspect as $EL^1 = (0.5, 0.6, 0.4)$ and $EL^2 = (0.6, 0.4, 0.4)$, respectively. Furthermore, the original HFDMs containing the judgments of projects with respect to the aspects are:

$$H^1 = \begin{pmatrix} \{0.6, 0.7\} & \{0.7\} & \{0.4, 0.6\} \\ \{0.6\} & \{0.6, 0.7\} & \{0.4\} \\ \{0.6\} & \{0.5\} & \{0.5, 0.7\} \end{pmatrix},$$

and

$$H^2 = \begin{pmatrix} \{0.7\} & \{0.5\} & \{0.3, 0.4, 0.5\} \\ \{0.5, 0.7\} & \{0.6\} & \{0.3\} \\ \{0.6, 0.8\} & \{0.4\} & \{0.5\} \end{pmatrix}.$$

4.6.2 Decision-Making Process

Based on the above information, the decision making process addressing by the proposed method can be summarized [12].

(1) Construct the corresponding HFPMs. The parameters in the value function are determined as $\alpha = \beta = 0.88$ and $\lambda = 2.25$ [6], the HFPMs can be obtained as:

$$\widetilde{H}^1 = \begin{pmatrix} \{0.2426, 0.4465\} & \{0.2952\} & \{0.3803\} \\ \{0.2426\} & \{0.2952\} & \{0\} \\ \{0.2426\} & \{0.2205\} & \{0.2066, 0.5434\} \end{pmatrix}$$

and

$$\widetilde{H}^2 = \begin{pmatrix} \{0.2952\} & \{0.2066\} & \{0.2411, 0.2066\} \\ \{0.2205, 0.2952\} & \{0.3803\} & \{0.2411\} \\ \{0.5434\} & \{0\} & \{0.2066\} \end{pmatrix}$$

(2) Calculate the hesitant decision energy matrix and the quality matrix for each decision maker by Eqs. (4.6) and (4.7):

(a) *Hesitant decision energy matrices*

$$E^1 = \begin{pmatrix} \{0.0970, 0.1786\} & \{0.0886\} & \{0.1141\} \\ \{0.0970\} & \{0.0886\} & \{0\} \\ \{0.0970\} & \{0.0662\} & \{0.0620, 0.1630\} \end{pmatrix}$$

and

$$E^2 = \begin{pmatrix} \{0.1181\} & \{0.0620\} & \{0.0620, 0.0723\} \\ \{0.0882, 0.1181\} & \{0.1141\} & \{0.0723\} \\ \{0.2174\} & \{0\} & \{0.0620\} \end{pmatrix}$$

(b) *Quality matrices*

$$q^1 = \begin{pmatrix} 0.9760 & 0.9723 & 0.9833 \\ 0.9693 & 0.9777 & 0.9337 \\ 0.9845 & 0.9733 & 0.9650 \end{pmatrix}$$

and

$$q^2 = \begin{pmatrix} 0.9683 & 0.9756 & 0.9808 \\ 0.9910 & 0.9833 & 0.9750 \\ 0.8932 & 0.8894 & 0.9514 \end{pmatrix}$$

where the mean hesitant decision potential of each hesitant decision energy matrix is computed as

$$\overline{h}_{E_1} = (\{0.0996, 0.1331\} \quad \{0.0663\} \quad \{0.0774, 0.1084\})^T$$

and

$$\overline{h}_{E_2} = (\{0.0879, 0.0849\} \quad \{0.0913, 0.1034\} \quad \{0.1106\})^T$$

(3) Get the hesitant decision exergy matrices of decision makers:

$$B^1 = \begin{pmatrix} \{0.0947, 0.1743\} & \{0.0861\} & \{0.1122\} \\ \{0.0940\} & \{0.0886\} & \{0\} \\ \{0.0955\} & \{0.0644\} & \{0.0598, 0.1573\} \end{pmatrix}$$

and

$$B^2 = \begin{pmatrix} \{0.1144\} & \{0.0605\} & \{0.0608, 0.0709\} \\ \{0.0874, 0.1170\} & \{0.1122\} & \{0.0705\} \\ \{0.1942\} & \{0\} & \{0.0590\} \end{pmatrix}$$

Then, by Eqs. (4.14)–(4.18), we get the hesitant decision entropy of each project as:

$$S_1 = \{0.0035, 0.0050\},\ S_2 = \{0.0022, 0.0023\},\ S_3 = \{0.0136, 0.0138\}.$$

Furthermore, the scores of $S_i (i = 1, 2, 3)$ are:

$$s(S_1) = 0.0043,\ s(S_2) = 0.0023,\ s(S_3) = 0.0137$$

The ranking of the projects in this emergency response is $P_2 \succ P_1 \succ P_3$, which shows that P_2 comprehensively performs best, and should be adopted in the emergency response.

4.6.3 *Comparing Case Results with Hesitant Fuzzy TOPSIS*

Ren et al. [12] furtherly used the hesitant fuzzy TOPSIS [24] to cope with the same emergency response selection problem and obtained the decision making results (as shown in Table 4.2).

In Table 4.2, different rankings of the projects are obtained by the proposed method and the hesitant fuzzy TOPSIS, which is caused by the different essences of the two methods. Ren et al. [12] pointed out the proposed method has relative advantages:

Table 4.2 The decision making results derived by hesitant fuzzy TOPSIS [12]

	d_i^+	d_i^-	c_i	Ranking
P_1	0.0650	0.1050	0.6176	1
P_2	0.0675	0.0525	0.4375	3
P_3	0.0700	0.0725	0.5089	2

(1) It depicts people's psychological behaviors, which is reasonable to be used in the situation where people are bounded rationality. This conforms to the realities, where humans cannot rationally think over things due to their different risk attitudes, incomplete information and other factors.
(2) It reflects not only the quantity of the decision making information as usual methods but also the quality. This character makes the utmost of the values and interior structure of the decision making information and is of great significance in handling uncertain and incomplete problems.

4.7 Summary

Considering the limitations of information and knowledge of decision makers and considering the difficulty in obtaining information, this chapter has introduced a thermodynamic method for MCDM based on prospect theory under a hesitant fuzzy environment. The main findings of the method can be briefly presented as follows [12]:

(1) Taking the advantages of hesitant fuzzy information to represent the fuzziness of the objects and the hesitant thoughts of decision makers.
(2) Introducing the negative exponential function into the prospect theory to portray decision makers' psychological behaviors during the decision making process.
(3) Applying the thermodynamics parameters to simultaneously take into account the quantity and the quality of the HFSs to make the best of the information.
(4) Designing experiments and utilizing the nonparametric test to demonstrate the validation of the proposed method.
(5) Solving the emergency response selection in the accident of firing and exploding at Port Group in Tianjin Binhai New Area to illustrate the proposed method.

References

1. Chen, N., Xu, Z.S., Xia, M.M.: Correlation coefficients of hesitant fuzzy sets and their applications to clustering analysis. Appl. Math. Model. **37**, 2197–2211 (2013)
2. Chiclana, F., Tapia García, J.M., del Moral, M.J., Herrera-Viedma, E.: A statistical comparative study of different similarity measures of consensus in group decision making. Inf. Sci. **221**, 110–123 (2013)
3. Dong, Y.C., Zhang, H.J., Herrera-Viedma, E.: Consensus reaching model in the complex and dynamic MAGDM problem. Knowl.-Based Syst. **106**, 206–219 (2016)
4. Fan, Z.P., Zhang, X., Chen, F.D., Liu, Y.: Multiple attribute decision making considering aspiration-levels: a method based on prospect theory. Comput. Ind. Eng. **65**, 341–350 (2013)
5. Gomes, L., Lima, M.: TODIM: basics and application to multi-criteria ranking of projects with environmental impacts. Found. Comput. Decis. Sci. **16**, 113–127 (1992)

6. Kahneman, D., Tversky, A.: Prospect theory: an analysis of decision under risk. Econometrica **47**(2), 263–292 (1979)
7. Lehmann, E.: Nonparametrics: Statistical Methods Based on Ranks. Holden-Day, San Francisco (1975)
8. Liao, H.C., Xu, Z.S., Xia, M.M.: Multiplicative consistency of hesitant fuzzy preference relation and its application in group decision making. Int. J. Inf. Technol. Decis. Mak. **13**(01), 47–76 (2014)
9. Liao, H.C., Xu, Z.S., Zeng, X.J.: Distance and similarity measures for hesitant fuzzy linguistic term sets and their application in multi-criteria decision making. Inf. Sci. **271**, 125–142 (2014)
10. Neumann, J.V., Morgenstern, O.: Theory of Games and Economic Behavior: 60th Anniversary Commemorative Edition. Princeton University Press, Princeton, New Jersey (2007)
11. Pérez, I.J., Cabrerizo, F.J. Herrera-Viedma, E.: A mobile decision support system for dynamic group decision making problems. IEEE Trans. Syst., Man Cybern. Part A: Syst. Hum., **40**, 1244–1256 (2010)
12. Ren, P.J., Xu, Z.S., Hao, Z.N.: Hesitant fuzzy thermodynamic method for emergency decision making based on prospect theory. IEEE Trans. Cybern. **47**(9), 2531–2543 (2017)
13. Saaty, T.L.: Rank from comparisons and from ratings in the analytic hierarchy/network processes. Eur. J. Oper. Res. **168**, 557–570 (2006)
14. Saaty, T.L., Shang, J.S.: An innovative orders-of-magnitude approach to AHP-based mutli-criteria decision making: prioritizing divergent intangible humane acts. Eur. J. Oper. Res. **214**, 703–715 (2011)
15. Torra, V.: Hesitant fuzzy sets. Int. J. Intell. Syst. **25**(6), 529–539 (2010)
16. Tversky, A., Kahneman, D.: Advances in prospect theory: cumulative representation of uncertainty. J. Risk Uncertain. **5**, 297–323 (1992)
17. Vroom, V.H.: Work and Motivation. England, Wiley, Oxford (1964)
18. Wang, J.Q. Sun, T.: Fuzzy multiple criteria decision making method based on prospect theory. In: Conference: Information Management, Innovation Management and Industrial Engineering (2008).
19. Wilcoxon, F.: Individual comparisons by ranking methods. Biometrics Bull. **1**, 80–83 (1945)
20. Xia, M.M., Xu, Z.S.: Hesitant fuzzy information aggregation in decision making. Int. J. Approximate Reasoning **52**, 395–407 (2011)
21. Xu, Z.S.: On multi-period multi-attribute decision making. Knowl.-Based Syst. **21**, 164–171 (2008)
22. Xu, Z.S.: Approaches to multistage multi-attribute group decision making. Int. J. Inf. Technol. Decis. Mak. **10**, 121–146 (2011)
23. Xu, Z.S., Xia, M.M.: Distance and similarity measures for hesitant fuzzy sets. Inf. Sci. **181**, 2128–2138 (2011)
24. Xu, Z.S., Zhang, X.L.: Hesitant fuzzy multi-attribute decision making based on TOPSIS with incomplete weight information. Knowl.-Based Syst. **52**, 53–64 (2013)
25. Zhang, X.L., Xu, Z.S.: The TODIM analysis approach based on novel measured functions under hesitant fuzzy environment. Knowl.-Based Syst. **61**, 48–58 (2013)

Chapter 5
A Thermodynamic Method for Heterogeneous Decision Making Based on Confidence Level

A decision making problem may contain both subjective criteria and objective criteria. For example, when investing in a company, a capitalist may consider the criteria of asset-liability ratio, annual profit, corporate image, business standing, enterprise development plan, etc. To solve the problem of involving the subjective and objective criteria, it is necessary to develop decision making methods under the heterogeneous environment.

With the development of human being's activities, as an essential branch of management, decision making has been applied in the fields of social economy [4, 7, 9, 13], military [1, 18, 19], etc. Initially, the classical decision making theory was established based on the expected utility theory [14], and Decision makers determine the most desirable alternative by evaluating the potential income and losses of all alternatives. However, the complexity and uncertainty of the modern decision making environment cause that Decision makers are challenging to assess alternatives accurately. To this end, decision making with fuzzy information has received significant development in the last decade. Linguistic term (LT) [24], as a type according with people's expression habits, can straightway portray Decision makers' judgments. It is noted that when Decision makers face a lot of uncertain factors, they might evaluate the alternatives with several possible linguistic terms. For example, compared by "good" or "very good", sometimes "between good and very good" is more in line with Decision makers' opinions. Considering this point, the hesitant fuzzy linguistic term set (HFLTS) [17], which represents the performance of an alternative through continuous LTs, has been widely used in the decision making field.

As verified in the previous chapters, making full use of decision making information has critical roles in the effectiveness of decision making. Based on the above analysis, the chapter will present a thermodynamic method for the decision making situation with real numbers (RNs), utility values (UVs), LTs, and HFLTSs developed by Ren et al. [16]. The method furtherly considers that different types of judgments represent Decision makers' different understanding of the alternatives and brings in

P. Ren and Z. Xu, *Decision-Making Analyses with Thermodynamic Parameters and Hesitant Fuzzy Linguistic Preference Relations*, Studies in Fuzziness and Soft Computing 409, https://doi.org/10.1007/978-3-030-73253-0_5

a confident degree to construct a weight adjustment process to improve the method's scientificity.

5.1 Linguistic Information

Zadeh [24] firstly introduced the concept of a linguistic term set (LTS), and represented it as $S = \{s_\alpha | \alpha = 0, 1, \ldots, 2\tau\}$, where τ is a positive number and s_α is a LT. For any two LTs s_α and s_β, operational laws of them are [22, 23]:

1. Scalar multiplication: $\lambda s_\alpha = s_{\lambda\alpha}$ $(\lambda \geq 0)$;
2. Addition: $s_\alpha \oplus s_\beta = s_{\alpha+\beta}$;
3. Subtraction: $s_\alpha \ominus s_\beta = s_{\alpha-\beta}$ for $\alpha > \beta$;
4. Negation operation: $Neg(s_\alpha) = s_{2\tau-\alpha}$.

Definition 5.1 ([22]). Let s_α and s_β be two LTs, then the Hamming distance of them can be calculated by

$$d(s_\alpha, s_\beta) = \frac{|s_\alpha - s_\beta|}{2\tau} \tag{5.1}$$

Based on a LTS $S = \{s_\theta | \theta = 0, 1, \ldots, 2\tau\}$ (τ is a positive number), a HFLTS [17] was defined to describe Decision makers' judgments towards objectives with several consecutive linguistic terms in S. Mathematically, it is noted as $b = \{\langle x_i, b(x_i)\rangle | x_i \in X\}$, where $b(x_i)$ is called a hesitant fuzzy linguistic element (HFLE) [10].

To ensure the operations of HFLTSs, Zhu and Xu [26] proposed a method to add some elements in the shorter one. If the shorter HFLTS is expressed as $b = \{b^l | l = 1, 2, ..., L\}$, then the added elements can be determined by $b = \eta\ (\max_l b^l) \oplus (1-\eta)\ (\min_l b^l)$ $(0 \leq \eta \leq 1)$.

Suppose that $b_\alpha = \{b^l_\alpha | l = 1, 2, \ldots, L\}$ and $b_\beta = \{b^l_\beta | l = 1, 2, \ldots, L\}$ are two HFLTSs with the same length, then the operations of them are [16, 26]:

1. Scalar multiplication: $\lambda b_\alpha = \cup_{b_\alpha^{\sigma(l)} \in b_\alpha} \{\lambda b_\alpha^{\sigma(l)}\}$ $(\lambda \geq 0)$;
2. Addition: $b_\alpha \oplus b_\beta = \cup_{\substack{b_\alpha^{\sigma(l)} \in b_\alpha \\ b_\beta^{\sigma(l)} \in b_\beta}} \{b_\alpha^{\sigma(l)} \oplus b_\beta^{\sigma(l)}\}$;
3. Subtraction: $b_\alpha \ominus b_\beta = \{I^{-1}(I(b_\alpha^{\sigma(l)}) - I(b_\beta^{\sigma(l)})) | l = 1, 2, ..., L\}$ for $I(b_\alpha^{\sigma(l)}) > I(b_\beta^{\sigma(l)})$;

where $b_\alpha^{\sigma(l)}$ and $b_\beta^{\sigma(l)}$ are the lth largest element in b_α and b_β, and $I(b)$ is the ordered lower index set of $b = \{b^l | l = 1, 2, \ldots, L\}$.

Definition 5.2 ([11]). Let $b_\alpha = \{b_\alpha^l | l = 1, 2, \ldots, L\}$ and $b_\beta = \{b_\beta^l | l = 1, 2, \ldots, L\}$ be two HFLTSs, then the Hamming distance of them is

$$d(b_\alpha, b_\beta) = \frac{1}{L}\sum_{l=1}^{L} \frac{\left| I\left(b_\alpha^{\sigma(l)}\right) - I\left(b_\beta^{\sigma(l)}\right)\right|}{2\tau} \tag{5.2}$$

Furthermore, Zhang and Wu [25] proposed a score function for a HFLTS $b = \{b^l | l = 1, 2, \ldots, L\}$ as

$$s(b) = \frac{1}{2\tau z}\sum_{l}^{z} I\left(b^l\right) \tag{5.3}$$

and subsequently introduced the following rules for comparing any two HFLTSs b_α and b_β:

1. If $s(b_\alpha) > s(b_\beta)$, then $b_\alpha > b_\beta$;
2. If $s(b_\alpha) = s(b_\beta)$, then $b_\alpha = b_\beta$.

5.2 Heterogeneous Thermodynamic Parameters

The section aims to present the concepts of decision thermodynamic parameters with heterogeneous information.

Firstly, a MCDM problem containing subjective and objective criteria can be briefly described [16]:

Suppose that a problem involves m alternatives $A = \{A_i | i = 1, 2, \ldots, m\}$, n subjective criteria $C = \{C_j | j = 1, 2, \ldots, n\}$, and o objective criteria $C = \{C_j | j = n+1, \ldots, n+o\}$. The Decision makers $D = \{D_k | k = 1, 2, \ldots, h\}$ assign the criteria weights as $w = (w_1, w_2, \ldots, w_{n+o})^T$, in which $w_j \in [0, 1]$ and $\sum_{j=1}^{n+o} w_j = 1$ for $j = 1, 2, \ldots, n+o$. The Decision makers use UVs, LTs or HFLTSs to evaluate the alternatives towards the subjective criteria, and RNs represent the objective performances of the alternatives $p_{ij} (i = 1, 2, \ldots, m; j = 1, 2, \ldots, n+o)$, then the heterogeneous decision matrix $R^k = (r_{ij}^k)_{m\times(n+o)}$ involving subjective and objective values can be presented as:

$$\text{for } k = 1, 2, \ldots, h \tag{5.4}$$

where uv_{ij}^k is an UV expressed by a crisp value, s_{ij}^k and $b_{ij}^k = \left\{(b_{ij}^k)^l \middle| l = 1, 2, \ldots, L_{b_{ij}^k}\right\}$ are a LT and a HFLTS based on the LTS $S = \{s_\theta | \theta = 0, 1, \ldots, 2\tau\}$, respectively.

The definitions of heterogeneous decision thermodynamic parameters can be given:

Definition 5.3 ([16]). In a decision making problem, the decision value on an alternative with respect to a criterion (subjective or objective) is defined as the heterogeneous decision potential (HDP), and the criterion weight is defined as the heterogeneous decision force (HDF), where the HDP can be presented by each form such as UV, LT, HFLTS, RN, etc., and the HDF is located in [0, 1].

Definition 5.4 ([16]). Heterogeneous decision energy (HDE) of an alternative represents its efficacy in the heterogeneous decision making system, which is the product of its HDP and the corresponding HDF.

Based on Definition 5.4, Ren et al. [16] provided a proposition to describe the HDE in detail:

Proposition 5.1 ([16]). For a heterogeneous decision matrix $R^k = (r_{ij}^k)_{m\times(n+o)}$ ($i \in \{1, 2, \ldots, m\}$, $j \in \{1, 2, \ldots, n+o\}$ and $k \in \{1, 2, \ldots, h\}$), where r_{ij}^k may be expressed as an UV uv_{ij}^k, a LT s_{ij}^k, a HFLTS b_{ij}^k or a RN p_{ij}, the HDE of the alternative A_i under the criterion C_j can be obtained by one of the following situations:

1. If the HDP is an UV uv_{ij}^k, then we call it utility energy, which can be calculated by $ue_{ij}^k = w_j uv_{ij}^k$;
2. If the HDP r_{ij}^k is a LT s_{ij}^k, then we call it linguistic energy, which can be calculated by $le_{ij}^k = w_j s_{ij}^k$;
3. If the HDP is a HFLTS b_{ij}^k, then we call it hesitant fuzzy linguistic energy, which can be calculated by $he_{ij}^k = w_j b_{ij}^k$;
4. If the HDP r_{ij}^k is a RN p_{ij}, then we call it real energy, which can be calculated by $re_{ij} = w_j p_{ij}$ [20].

The quality of heterogeneous information can be further defined as:

Definition 5.5 ([16]). Towards a criterion, the concentration degree of HDPs on an alternative is called its heterogeneous quality (HQ).

Proposition 5.2 ([16]). For a heterogeneous decision matrix $R^k = (r_{ij}^k)_{m\times(n+o)}$ ($i \in \{1, 2, \ldots, m\}$, $j \in \{1, 2, \ldots, n+o\}$ and $k \in \{1, 2, \ldots, h\}$), where r_{ij}^k may be expressed as an UV uv_{ij}^k, a LT s_{ij}^k, a HFLTS b_{ij}^k or a RN p_{ij}, then the HQ of a HDP r_{ij}^k ($k \in \{1, 2, \ldots, h\}$) is

1. If the HDP is an UV uv_{ij}^k, then the HQ is called utility quality, computed by $uq_{ij}^k = 1 - \frac{\left|uv_{ij}^k - \overline{uv}_{ij}\right|}{\overline{uv}_{ij}}$;
2. If the HDP is a LT s_{ij}^k, then the HQ is called linguistic quality, computed by $lq_{ij}^k = 1 - d(s_{ij}^k, \overline{s}_{ij})$;
3. If the HDP is a HFLTS b_{ij}^k, then HQ is called hesitant fuzzy linguistic quality, computed by $hq_{ij}^k = 1 - d(b_{ij}^k, \overline{b}_{ij})$;
4. If the HDP r_{ij}^k is a RN, then $r_{ij}^k = r_{ij}^{k+1}$ for all $k \in \{1, 2, \ldots, h\}$, and the real quality is $rq_{ij} = 1$;

where $\overline{uv}_{ij}, \overline{s}_{ij}, \overline{b}_{ij}$ are the central values of the HDPs, $d(s_{ij}^k, \overline{s}_{ij})$ and $d(b_{ij}^k, \overline{b}_{ij})$ are the distance measures of linguistic information and hesitant fuzzy linguistic information, respectively.

Similar to Chaps. 3 and 4, heterogeneous decision exergy (HDEX) and heterogeneous decision entropy (HDEN) can be introduced.

Definition 5.6 ([16]). HDEX is defined as the maximum centralized effect that a HDP possesses. It can be calculated by the product of HDE and HQ of each alternative.

Proposition 5.3 ([16]). For a heterogeneous decision matrix $R^k = (r_{ij}^k)_{m\times(n+o)}$ ($i \in \{1, 2, \ldots, m\}, j \in \{1, 2, \ldots, n+o\}$ and $k \in \{1, 2, \ldots, h\}$), where r_{ij}^k is an UV uv_{ij}^k, a LT s_{ij}^k, a HFLTS b_{ij}^k or a RN p_{ij}, then the HDEX of a HDP r_{ij}^k ($k \in \{1, 2, \ldots, h\}$) is

1. For an UV uv_{ij}^k, its utility exergy $ux_{ij}^k = uq_{ij}^k \cdot ue_{ij}^k$;
2. For a LT s_{ij}^k, its linguistic exergy $lx_{ij}^k = lq_{ij}^k \cdot le_{ij}^k$;
3. For a HFLTS b_{ij}^k, its hesitant fuzzy linguistic exergy $hx_{ij}^k = hq_{ij}^k \cdot he_{ij}^k$;
4. For a RN p_{ij}, its real exergy $rx_{ij} = rq_{ij} \cdot re_{ij}$ [20].

Definition 5.7 ([16]). HDEN is defined as the maximum decentralized effect that a HDP possesses. It is the difference value between the HDE and HDEX of an alternative.

Proposition 5.4 ([16]). For a heterogeneous decision matrix $R^k = (r_{ij}^k)_{m\times(n+o)}$ ($i \in \{1, 2, \ldots, m\}, j \in \{1, 2, \ldots, n+o\}$ and $k \in \{1, 2, \ldots, h\}$), where r_{ij}^k is presented by an UV uv_{ij}^k, a LT s_{ij}^k, a HFLTS b_{ij}^k or a RN p_{ij}, then the HDEN of any HDP r_{ij}^k ($k \in \{1, 2, \ldots, h\}$) is

1. Utility entropy of an UV uv_{ij}^k is $un_{ij}^k = ue_{ij}^k - ux_{ij}^k$;
2. Linguistic entropy of a LT s_{ij}^k is $\ln_{ij}^k = le_{ij}^k \ominus lx_{ij}^k$;
3. Hesitant fuzzy linguistic entropy of a HFLTS b_{ij}^k is $hn_{ij}^k = he_{ij}^k \ominus hx_{ij}^k$;
4. Real entropy of a RN p_{ij} is $rn_{ij} = re_{ij} - rx_{ij}$ [20].

5.3 Weights Modification Process

Considering that people's fuzzy mind always manifests as ambiguous expressions, this section introduces a method developed by Ren et al. [16] to modify the criteria weights to relatively decrease the importance of the criteria corresponding to ambiguous expressions and increase the importance of the criteria corresponding to accurate expressions.

Definition 5.8 ([16]). The confidence level of a decision-maker with respect to his/her given HFLTS $b = \left\{b^l | l = 1, 2, \ldots, L_b\right\}$ is defined to measure its dependability in the decision making process, which is calculated by

$$\xi = \frac{2}{L_b + 1} \tag{5.5}$$

Ren et al. [16] noted that the confidence level is proposed based on the definition of relative standard deviation in mathematics (statistics). Then, they introduced the following technique to modify the criteria weights.

Definition 5.9 ([16]). Suppose that an uncertain decision value corresponds to the original weight w, and its confidence level is calculated as ξ, then its weight can be modified by

$$\overline{w} = w \cdot f(\xi) \tag{5.6}$$

where $f(\xi)$ is a function of the confidence level ξ.

Proposition 5.5 ([16]). The form of $f(\xi)$ can be determined by a decision-maker's sensitive attitude towards the accuracy of decision value, specifically

1. If the decision-maker is neutrally sensitive to the accuracy, then $f(\xi)$ should be a linear function, $f''(\xi) = 0$;
2. If the decision-maker is sensitive to the accuracy, then $f(\xi)$ should be a locally convex function, $f''(\xi) > 0$;
3. If the decision-maker is insensitive to the accuracy, then $f(\xi)$ should be a locally concave function, $f''(\xi) < 0$.

5.4 Decision-Making with Thermodynamic Parameters and Confidence Level Under Heterogeneous Environment

For a MCDM problem with m alternatives $A = \{A_i | i = 1, 2, \ldots, m\}$, n subjective criteria $C = \left\{C_j | j = 1, 2, \ldots, n\right\}$, and o objective criteria $C = \left\{C_j | j = n + 1, \ldots, n + o\right\}$, the Decision makers $D = \{D_k | k = 1, 2, ..., h\}$ assign the vectors of criteria weights as $w^k = (w_1^k, w_2^k, ..., w_{n+o}^k)^T$ for $k = 1, 2, ..., h$. RNs p_{ij} $(i = 1, 2, ..., m, \ \ j = 1, 2, ..., n)$ represent the performances of alternatives towards objective criteria, the Decision makers furtherly give the judgments on alternatives towards subjective criteria (expressed by UVs uv_{ij}^k, LTs s_{ij}^k, $C = \left\{C_j | j = n + 1, \ldots, n + o\right\}$ and HFLTSs b_{ij}^k for $i = 1, 2, ..., m$, $j = 1, 2, ..., n$ and $k = 1, 2, ..., h$), the heterogeneous decision matrices are denoted as $R^k = (r_{ij}^k)_{m \times (n+o)}$ $(k = 1, 2, ..., h)$. Suppose that the Decision makers' functions of the confidence levels according to their sensitive preferences is $f(\xi_{j*}^k)$, then the decision making process with thermodynamic parameters and confidence level under heterogeneous environment can be shown as follows:

Algorithm 5.1 ([16]). **Weights rectifying**

Step 1. For all heterogeneous decision matrices $R^k = (r_{ij}^k)_{m\times(n+o)}(i = 1, 2, ..., m$, $j = 1, 2, ..., n + o$ and $k = 1, 2, ..., h)$, we select the criteria C_j^k that correspond to the evaluations expressed by HFLTSs for $j = 1, 2, ..., n$ and $k = 1, 2, ..., h$, go the next step; If no criterion as such exist, then end this algorithm;

Step 2. Denote the selected criteria and their original weights as $C_{j^*}^k$ $(j = 1, 2, ..., n,\ k = 1, 2, ..., h)$ and $w_{j^*}^k$ $(j = 1, 2, ..., n,\ \ k = 1, 2, ..., h)$.

Step 3. Obtain the confidence levels of the Decision makers towards the selected criteria $C_{j^*}^k$ $(j = 1, 2, ..., n,\ \ k = 1, 2, ..., h)$ by

$$\xi_{j^*}^k = \frac{2}{L_{b_{j^*}^k} + 1} \tag{5.7}$$

where $L_{b_{j^*}^k}$ is the length of the normalized HFLTSs for the jth criterion given by the kth decision-maker; $k = 1, 2, ..., h$

Step 4. Calculate the modified weights of the criteria C_{j^*} $(j = 1, 2, ..., n)$, by

$$\bar{w}_{j^*}^k = w_{j^*}^k \cdot f\left(\xi_{j^*}^k\right) \tag{5.8}$$

Step 5. For the criteria expressed by UVs, LTs and RNs, we adjust their weights by

$$\overline{w}_j^k = w_j^k\left(1 + \sum_{j^*=1}^{n} \left|w_{j^*}^k - \overline{w}_{j^*}^k\right| / \left(1 - \sum_{j^*=1}^{n} \overline{w}_{j^*}^k\right)\right),$$
$$\text{for} \quad j = 1, 2, ..., n + o \quad \text{and} \quad k = 1, 2, ..., h \tag{5.9}$$

and then output the modified criteria weights $\overline{w}^k = (\overline{w}_1^k, \overline{w}_2^k, ..., \overline{w}_{n+o}^k)^T$ for $k = 1, 2, ..., h$.

Algorithm 5.2 ([16]). **Thermodynamic features processing**

Step 1. Obtain the HDE of each HDP by Proposition 5.1, and construct the heterogeneous energy matrices $HE^k = (he_{ij}^k)_{m\times(n+o)}$ of the heterogeneous decision matrices $R^k = (r_{ij}^k)_{m\times(n+o)}$ for $k = 1, 2, ..., h$;

Step 2. Calculate the heterogeneous quality matrices $HQ^k = (hq_{ij}^k)_{m\times(n+o)}$ of the heterogeneous decision matrices for $k = 1, 2, ..., h$, where hq_{ij}^k can be obtained by Proposition 5.2;

Step 3. Acquire the heterogeneous exergy matrices $HX^k = (hx_{ij}^k)_{m\times(n+o)}$ and the heterogeneous entropy matrices $HN^k = (hn_{ij}^k)_{m\times(n+o)}$ by Propositions 5.3 and 5.4 for $k = 1, 2, ..., h$, respectively;

Step 4. Synthesize all heterogeneous exergy matrices and heterogeneous entropy matrices into overall ones, denoted as $HX = (hx_{ij})_{m\times(n+o)}$ and $HN = (hn_{ij})_{m\times(n+o)}$, where

$$hx_{ij} = \frac{1}{h}\sum_{k=1}^{h} hx_{ij}^{k} \tag{5.10}$$

$$hn_{ij} = \frac{1}{h}\sum_{k=1}^{h} hn_{ij}^{k} \tag{5.11}$$

Considering that transforming heterogeneous data into the same type may lose some original information, Ren et al. [16] proposed to apply the TODIM [5] to directly handle the heterogeneous exergy matrices (the better performance of an alternative in heterogeneous exergy matrices indicates a higher ranking) by relatively comparing alternatives with respect to each criterion.

Algorithm 5.3 ([16]). **TODIM procedure for heterogeneous information with thermodynamic features**

Step 1. Normalize the elements in the heterogeneous exergy matrices that are expressed by UVs or RNs into 0–1 by

$$\overline{hx}_{ij} = \frac{hx_{ij}}{\max_{\substack{i\in\{1,2,...,m\}\\ j\in\{1,2,...,n+o\}}} \left(hx_{ij}\right)} \tag{5.12}$$

Step 2. Obtain the overall rectified weights $\overline{w} = (\overline{w}_1, \overline{w}_2, ..., \overline{w}_{n+o})^T$ based on the individual rectified weights $\overline{w}^k = (\overline{w}_1^k, \overline{w}_2^k, ..., \overline{w}_{n+o}^k)^T$ for $k = 1, 2, ..., h$ by

$$\overline{w}_j = \frac{1}{k}\sum_{k=1}^{h} \overline{w}_j^k \tag{5.13}$$

Step 3. Calculate the relative weight of the criterion C_j to the criterion C_r

$$w_{jr} = \overline{w}_r / \overline{w}_j \quad \text{for} \quad j, r = 1, 2, ..., n+o \tag{5.14}$$

Step 4. Get the dominance degree of the alternative A_i over the alternative A_t with respect to each criterion C_r by

1. If $\overline{hx}_{ij}$ is a UV or a RN, then

$$\delta_r(A_i, A_t)) = \begin{cases} \sqrt{\dfrac{w_{jr}\left(\overline{hx}_{ir} - \overline{hx}_{tr}\right)}{\sum_{r=1}^{n+o} w_{jr}}} & if\ \left(\overline{hx}_{ir} - \overline{hx}_{tr}\right) > 0 \\ -\dfrac{1}{\theta}\sqrt{\dfrac{\left(\sum_{r=1}^{n+o} w_{jr}\right)\left(\overline{hx}_{tr} - \overline{hx}_{ir}\right)}{w_{jr}}} & otherwise \end{cases} \tag{5.15}$$

2. If $\overline{hx}_{ij}$ is a LT or a HFLTS, then

$$\delta_r(A_i, A_t) = \begin{cases} \sqrt{\dfrac{w_{jr} d\left(\overline{hx}_{ir}, \overline{hx}_{tr}\right)}{\sum_{r=1}^{n+o} w_{jr}}} & if\ \overline{hx}_{ir} > \overline{hx}_{tr} \\ -\dfrac{1}{\theta}\sqrt{\dfrac{\left(\sum_{r=1}^{n+o} w_{jr}\right) d\left(\overline{hx}_{tr}, \overline{hx}_{ir}\right)}{w_{jr}}} & otherwise \end{cases} \tag{5.16}$$

where $d(\overline{hx}_{ir}, \overline{hx}_{tr})$ is the distance between $\overline{hx}_{ir}$ and $\overline{hx}_{tr}$;

Step 5. Calculate the dominance degree of the alternative A_i over the alternative A_t by

$$\Delta(A_i, A_t) = \sum_{r=1}^{n+o} \delta_r(A_i, A_t) \tag{5.17}$$

Step 6. Obtain the global value of each alternative by

$$\gamma_i = \frac{\sum_{t=1}^{m} \Delta(A_i, A_t) - \min_i\left(\sum_{t=1}^{m} \Delta(A_i, A_t)\right)}{\max_i\left(\sum_{t=1}^{m} \Delta(A_i, A_t)\right) - \min_i\left(\sum_{t=1}^{m} \Delta(A_i, A_t)\right)} \quad \text{for} \quad i = 1, 2, \ldots, m \tag{5.18}$$

rank alternatives according to their global values.

5.5 Discussions

The section presents the experiments produced by Ren et al. [16] that discuss the impacts of Decision makers' sensitive attitudes on decision results.

Suppose that for a MCDM problem with four alternatives and four criteria (including an objective criterion and three subjective criteria), three Decision makers are invited to give their judgments to the alternatives with respect to each subjective criterion by UVs, LTs and HFLTSs, respectively, where LTs and HFLTSs are based on $S = \{s_\theta | \theta = 0, 1, \ldots, 2\tau\}$ (τ is a positive number). Then, randomly generate three decision matrices that represent three Decision makers' judgments and three vectors of criteria weights, and address the weight vectors by the following situations [16]:

Situation 1. For the column that is evaluated by HFLTSs, suppose that Decision makers are neutrally sensitive towards the accuracy of decision information, then set the function of the confidence level as $f(\xi) = \xi$ to modify the corresponding weight;

Situation 2. For the column that is evaluated by HFLTSs, suppose that Decision makers are sensitive towards the accuracy of decision information, then set the function of the confidence level as $f(\xi) = e^{\xi} - 1$ to modify the corresponding weight;

Situation 3. For the column that is evaluated by HFLTSs, suppose that Decision makers are insensitive towards the accuracy of decision information, then set the function of the confidence level as $f(\xi) = \ln(\xi + 1)$ to modify the corresponding weight;

Situation 4. For the column that is evaluated by HFLTSs, we reserve its original weight.

Substitute the weight vectors obtained under the above four situations into Algorithms 5.2 and 5.3, and get the corresponding four rankings of alternatives and record the optimal selection of each ranking. Repeat this procedure 1000 times, and list four optimal selection rankings towards the four situations.

Let the decision matrices be with five alternatives and five criteria, six alternatives and six criteria, seven alternatives and seven criteria, eight alternatives and eight criteria, nine alternatives and nine criteria, ten alternatives, and ten criteria, conduct the above process and get the rankings with different situations under each order of decision matrices.

Ren et al. [16] indicated that a way to analyze the impacts of sensitive attitudes on decision results is to compare the rankings with the original weights and the rectified weights towards different sensitive attitudes. Considering that Pearson product-moment correlation coefficient (PPMCC) is an efficient technique to explore the relationship between two sequences of data without centralization, they proposed to apply it to test the correlations of any two ranking sequences derived from the above random process and get the results in Table 5.1.

Some conclusions can be obtained [16]:

1. The correlation coefficients in Table 5.1 are all significant. The test shows that the decision results obtained through the original weights and the modified weights with sensitive attitude/neutrally sensitive attitude are correlative.
2. The decision results obtained by the original weights and the modified weights with insensitive attitudes can completely fit with each other because the correlation coefficients are all 1. It verifies that Decision makers' insensitive attitude

Table 5.1 The correlations between the rankings with the original weights and the rectified weights with different sensitive attitudes [16]

Correlation coefficients	Decision matrices 4*4	Decision matrices 5*5	Decision matrices 6*6	Decision matrices 7*7
Neutrally sensitive (measured by function $f(\xi) = \xi$)	0.950 Sig. 0.000 (2-tailed)	0.995 Sig. 0.000 (2-tailed)	0.997 Sig. 0.000 (2-tailed)	1.000 Sig. 0.000 (2-tailed)
Sensitive (measured by function $f(\xi) = e^{\xi} - 1$)	0.895 Sig. 0.000 (2-tailed)	0.916 Sig. 0.000 (2-tailed)	0.915 Sig. 0.000 (2-tailed)	0.914 Sig. 0.000 (2-tailed)
Insensitive [measured by function $f(\xi) = \ln(\xi + 1)$]	1.000 Sig. 0.000 (2-tailed)	1.000 Sig. 0.000 (2-tailed)	1.000 Sig. 0.000 (2-tailed)	1.000 Sig. 0.000 (2-tailed)
Correlation coefficients	Decision matrices 8*8	Decision matrices 9*9	Decision matrices 10*10	
Neutrally sensitive (measured by function $f(\xi) = \xi$)	1.000 Sig. 0.000 (2-tailed)	1.000 Sig. 0.000 (2-tailed)	1.000 Sig. 0.000 (2-tailed)	
Sensitive (measured by function $f(\xi) = e^{\xi} - 1$)	0.894 Sig. 0.000 (2-tailed)	0.906 Sig. 0.000 (2-tailed)	0.921 Sig. 0.000 (2-tailed)	
Insensitive [measured by function $f(\xi) = \ln(\xi + 1)$]	1.000 Sig. 0.000 (2-tailed)	1.000 Sig. 0.000 (2-tailed)	1.000 Sig. 0.000 (2-tailed)	

towards the accuracy of decision information does not influence the final optimal selections.

3. Generally, the results in Table 5.1 accord with Decision makers' behaviors, which indicates that the influence of the modified weights decreases with the decreasing sensitivity of Decision makers.
4. Decision results obtained by the modified weights with neutrally sensitive/sensitive attitudes can change the optimal selections.

To further show the impacts of different sensitive attitudes towards uncertain decision information on optimal selections, Ren et al. [16] compared the optimal selections with the modified weights and the optimal selections with the original weights and presented the differences in Fig. 5.1.

Ren et al. [16] summarized the observations by Fig. 5.1:

1. With the increasing order of the decision matrix, the number of different optimal selections between the modified weights (neutrally sensitive attitude or sensitive attitude) and the original weights generally decreases.

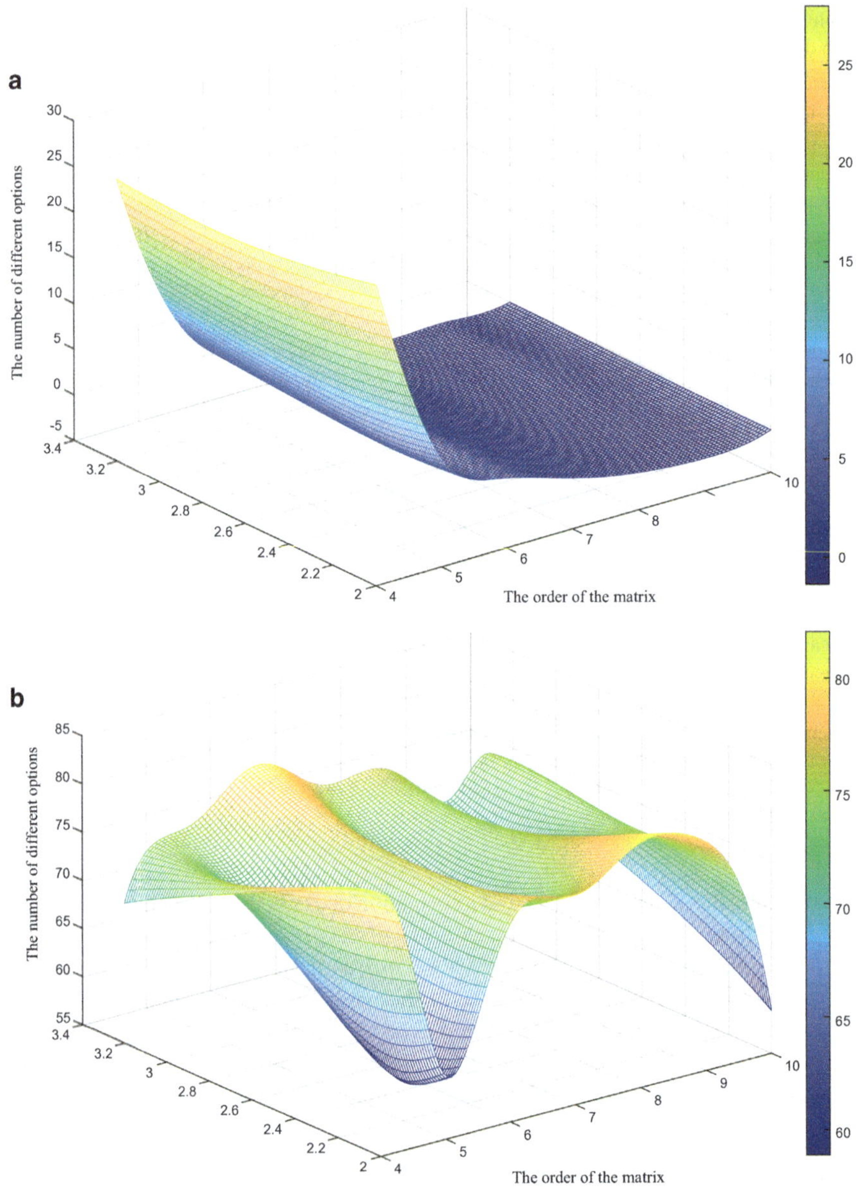

Fig. 5.1 Different optimal selections with the modified weights and the original weights [16]. **a** The number of different selections derived by the modified weights with neutrally sensitive attitude and the original weights **b** The number of different selections derived by the modified weights with sensitive attitudes and the original weights

2. If Decision makers' attitudes towards uncertain information are neutrally sensitive, then the optimal selection is difficult to be impacted by the modified weights when the order of decision matrix is equal to or greater than seven.

5.6 A Case Study on Green Supplier Selection Under a Low-Carbon Economy

5.6.1 Description of the Case

Ren et al. [16] used the proposed heterogeneous decision making method to find a green supplier promoting long-term cooperation under a low-carbon economy. Firstly, the background of the case was introduced as follows [16]:

In recent years, global warming destructs biodiversity, reduces crops, increases diseases, and threatens humanity's survival and sustainable development [3]. Because the primary cause of global warming is the massive emissions of greenhouse gas and carbon dioxide is the major kind of greenhouse gas, numerous countries have focused on the issue of carbon emissions, and the concept of low-carbon economy has been proposed since Copenhagen summit of 2009.

Laws have been promulgated to address this issue, such as imposing the carbon tariff on imported goods in the developed countries, implementing carbon tax in Singapore, introducing carbon tax policy in China, etc. Besides, the low-carbon wave has been brought force in commercial activity. Some corporations have considered low-carbon in their industrial chains. For example, Dell has been requiring the major suppliers to report their carbon emissions and proposing its carbon emissions requirements since 2007; Wal-Mart has been demanding all product suppliers to establish their low-carbon certification and reporting system.

As low-carbon has become the inevitable trend in commercial activity, it is necessary to consider this factor in a supply chain, which is beneficial for a corporation's long-run development. To this end, selecting an appropriate green supplier is worth being discussed in the corporation's supply chain management.

5.6.2 Indicators Selection

Noci [15] proposed the indicators for evaluating green suppliers can be decided as green competitiveness, environmental management efficiency, green image, and life cycle. Later on, Humphreys et al. [6] provided a set of indicators for evaluating the environmental performance of a supplier, and further proposed combining the qualitative indicators and the quantitative indicators to judge suppliers. Wang [21] summarized that it is valid to measure environmental performance by green image, contamination control, environmental management system, green design, green capability, product recycling, and pollution treatment cost. Moreover, indicator systems for different industries, such as the electronics industry [8, 12], manufacturing industry [2], were built to address the corresponding issue.

Based on the above literature, Ren et al. [16] pointed out that the indicators can be chosen as (1) *Quality*. It includes the certificates of quality, the efficient ways to control non-conforming products, and the measures to rectify or recall non-conforming products; (2) *Delivery reliability*. It is the ratio of the orders delivered and total orders; (3) *Flexibility*. It includes the response capacity to react to the particular requirements on products, the diversity of products it provides, and the ability to deal with information sharing; (4) *Green image*. It presents a supplier's responsibility on social benefits, enterprise and society's long-term development; (5) *Technical and green research and development ability (R&D ability)*. It measures a supplier's green innovation capacity.

5.6.3 Decision-Making Process

Ren et al. [16] supposed that there are four suppliers for a manufacturing corporation to be chosen. A leadership, including three division heads from the purchasing, operation, and planning departments, participates in determining a supplier to cooperate. Denote the above five criteria as C_1, C_2, C_3, C_4 and C_5, suppose that the three heads assign the weights of criteria as $w_1 = (0.19,\ 0.25,\ 0.16,\ 0.20,\ 0.20)^T$, $w_2 = (0.20, 0.22,\ 0.17,\ 0.19,\ 0.22)^T$ and $w_3 = (0.18, 0.22, 0.20, 0.19,\ 0.21)^T$, and give their judgments as: where the judgments with respect to C_2 are expressed by RNs, with respect to C_5 are expressed by UVs, with respect to C_1 and C_4 are expressed by LTs, with respect to C_3 are expressed by HFLTSs. LTs and HFLTSs in Table 5.2 are assigned based on the linguistic term set:

$$S = \left\{ \begin{array}{l} s_0 : \text{extremely poor},\ s_1 : \text{very poor},\ s_2 : \text{poor}, \\ s_3 : \text{slightly poor},\ s_4 : \text{fair},\ s_5 : \text{slightly good}, \\ s_6 : \text{good},\ s_7 : \text{very good},\ s_8 : \text{extremely good} \end{array} \right\}.$$

The decision making process can be listed as follows [16]:

1. Suppose that the function $f(\xi) = e^{\xi} - 1$ can represent Decision makers' sensitive preferences, then the modified weights can be obtained as $\overline{w}_1 = (0.20, 0.27, 0.1, 0.21, 0.22)^T$, $\overline{w}_2 = (0.22,\ 0.23,\ 0.11,\ 0.20,\ 0.24)^T$, and $\overline{w}_3 = (0.18, 0.22, 0.19, 0.19,\ 0.22)^T$.
2. By Algorithm 5.2, the heterogenous exergy matrix can be got (as shown in Table 5.3).
3. Apply Algorithm 5.3 to derive the final ranking of the suppliers. The results are listed in Table 5.4.

Table 5.4 suggests that the ranking of the suppliers is determined as:

Table 5.2 Decision matrices of the green supplier selection problem [16]

	C_1	C_2	C_3	C_4	C_5
Head 1					
Supplier 1	s_6	0.95	$\{s_5, s_6, s_7\}$	s_6	7
Supplier 2	s_6	0.90	$\{s_5, s_6\}$	s_7	9
Supplier 3	s_4	0.92	$\{s_6, s_7\}$	s_7	10
Supplier 4	s_5	0.88	$\{s_5, s_6, s_7\}$	s_5	10
Head 2					
Supplier 1	s_6	0.95	$\{s_5, s_6\}$	s_5	9
Supplier 2	s_7	0.90	$\{s_4, s_5, s_6\}$	s_7	9
Supplier 3	s_4	0.92	$\{s_5\}$	s_6	11
Supplier 4	s_5	0.88	$\{s_6, s_7\}$	s_6	10
Head 3					
Supplier 1	s_7	0.95	$\{s_7\}$	s_6	8
Supplier 2	s_7	0.90	$\{s_4, s_5\}$	s_6	9
Supplier 3	s_4	0.92	$\{s_4, s_5\}$	s_6	11
Supplier 4	s_6	0.88	$\{s_6, s_7\}$	s_6	11

Table 5.3 Heterogenous exergy matrix of the green supplier selection problem [16]

	C_1	C_2	C_3	C_4	C_5
Supplier 1	$s_{1.25}$	0.23	$\{s_{0.76}, s_{0.79}, s_{0.86}\}$	$s_{1.12}$	1.59
Supplier 2	$s_{1.31}$	0.23	$\{s_{0.56}, s_{0.59}, s_{0.73}\}$	$s_{1.32}$	1.96
Supplier 3	$s_{0.71}$	0.22	$\{s_{0.63}, s_{0.72}\}$	$s_{1.25}$	2.27
Supplier 4	$s_{1.06}$	0.21	$\{s_{0.74}, s_{0.78}, s_{0.91}\}$	$s_{1.12}$	2.25

Table 5.4 Decision results of the green supplier selection problem [16]

Dominance degree	Supplier 1	Supplier 2	Supplier 3	Supplier 4	Global value	Ranking
Supplier 1	0	−0.47	−0.31	−0.30	*0*	*4*
Supplier 2	0.18	0	−0.21	−0.23	*1*	*1*
Supplier 3	−0.11	−0.37	0	−0.12	*0.59*	*2*
Supplier 4	−0.09	−0.30	−0.22	0	*0.57*	*3*

$$Supplier\ 2 \succ Supplier\ 3 \succ Supplier\ 4 \succ Supplier\ 1,$$

indicating that the leadership should choose Suppliers 2 for the corporation [16].

5.7 Summary

Since that Decision makers may have different cognitions to different criteria, different information types may exist in the decision making process. Under such an environment, the chapter has presented a decision making method with thermodynamic parameters and confidence level under heterogeneous environment, which contributes to [16].

1. Allowing Decision makers to utilize the natural linguistic representation besides the classical types of real numbers and utility values.
2. Reducing the impact of fuzzy thinking during the decision making process by decreasing the weight of the criterion with relatively vague judgment.
3. Increasing the rationality of the decision results by providing a technique for combining numerical values and data distribution of judgments during the decision making process.
4. Avoiding the information loss and distortion by introducing TODIM to solve the heterogeneous matrix with the thermodynamic feature.
5. Verifying the reasonability of the proposed method by making simulations for discussing how Decision makers' sensitive attitudes towards uncertainty impact decision results.

References

1. Burton, O.V.: Lincoln and leadership: Military, political, and religious decision making. J. Mil. Hist. **78**, 1123–1124 (2014)
2. Büyüközkan, G., Çifçi, G.: A novel fuzzy multi-criteria decision framework for sustainable supplier selection with incomplete information. Comput. Ind. **62**(2), 164–174 (2011)
3. Cheng, S.W.: New energy and low carbon economy. Manage. Rev. **22**, 4–8 (2010)
4. Farahani, R.Z., Asgari, N.: Combination of MCDM and covering techniques in a hierarchical model for facility location: a case study. Eur. J. Oper. Res. **176**, 1839–1858 (2007)
5. Gomes, L., Lima, M.: TODIM: basics and application to multi-criteria ranking of projects with environmental impacts. Found. Comput. Decis. Sci. **16**, 113–127 (1992)
6. Humphreys, P., McIvor, R., Chan, F.: Using case-based reasoning to evaluate supplier environmental management performance. Expert Syst. Appl. **25**(2), 141–153 (2003)
7. Johansson-Stenman, O.: Animal welfare and social decisions: Is it time to take bentham seriously. Ecol. Econ. **145**, 90–103 (2018)
8. Lee, A.H.I., Kang, H.Y., Hsu, C.F., Hung, H.C.: A green supplier selection model for high-tech industry. Expert Syst. Appl. **36**(4), 7917–7927 (2009)
9. Leiser, D., Azar, O.H.: Behavioral economics and decision making: Applying insights from psychology to understand how people make economic decisions. J. Econ. Psychol. **29**, 613–618 (2008)
10. Liao, H.C., Xu, Z.S.: Approaches to manage hesitant fuzzy linguistic information based on the cosine distance and similarity measures for HFLTSs and their application in qualitative decision making. Expert Syst. Appl. **42**, 5328–5336 (2015)
11. Liao, H.C., Xu, Z.S., Zeng, X.J.: Distance and similarity measures for hesitant fuzzy linguistic term sets and their application in multi-criteria decision making. Inf. Sci. **271**, 125–142 (2014)

12. Lu, L.Y.Y., Wu, C.H., Kuo, T.C.: Environmental principles applicable to green supplier evaluation by using multi-objective decision analysis. Int. J. Prod. Res. **45**, 4317–4331 (2007)
13. Mendoza, R.L.: Bringing the patient back in: behavioral decision making and choice in medical economics. J. Med. Econ. **21**(4), 313–317 (2018)
14. Neumann, J.V., Morgenstern, O.: Theory of games and economic behavior (60th Anniversary Commemorative Edition). Princeton University Press, Princeton, New Jersey (2007)
15. Noci, G.: Designing "green" vendor rating systems for the assessment of a suppliers environmental performance. Eur. J. Purchasing Supply Manage. **2**, 103–114 (1997)
16. Ren, P.J., Xu, Z.S., Verma, M., Zeng, X.J., Liao, H.C., Wang, X.X.: Heterogeneous group decision making with thermodynamical parameters and confidence level. Technical Report (2020)
17. Rodríguez, R.M., Martínez, L., Herrera, F.: Hesitant fuzzy linguistic terms sets for decision making. IEEE Trans. Fuzzy Syst. **20**(1), 109–119 (2012)
18. Salinas, G.: The decision to attack: military and intelligence cyber decision making. Perspect. Polit. **15**, 283–284 (2017)
19. Sánchez-Lozano, J.M., Serna, J., Dolón-Payán, A.: Evaluating military training aircrafts through the combination of multi-criteria decision making processes with fuzzy logic. A case study in the Spanish Air Force Academy. Aerosp. Sci. Technol. **42**, 58–65 (2018)
20. Verma, M., Rajasankar, J.: A thermodynamical approach towards group multi-criteria decision making (GMCDM) and its application to human resource selection. Appl. Soft Comput. **52**, 323–332 (2017)
21. Wang, X.X.: Review of researches on green supplier selection. Econ. Res. Guid. **4**, 194–196 (2017)
22. Xu, Z.S.: EOWA and EOWG operators for aggregating linguistic labels based on linguistic preference relations. Internat. J. Uncertain. Fuzziness Knowl.-Based Syst. **12**(06), 791–810 (2004)
23. Xu, Z.S.: Deviation measures of linguistic preference relations in group decision making. Omega **33**(3), 249–254 (2005)
24. Zadeh, L.A.: The concept of a linguistic variable and its application to approximate reasoning-Part I. Inf. Sci. **8**(3), 199–249 (1975)
25. Zhang, Z.M., Wu, C.: On the use of multiplicative consistency in hesitant fuzzy linguistic preference relations. Knowl.-Based Syst. **72**, 13–27 (2014)
26. Zhu, B., Xu, Z.S.: Consistency measures for hesitant fuzzy linguistic preference relations. IEEE Trans. Fuzzy Syst. **22**, 34–45 (2014)

Chapter 6
A Priority Programming Model for Hesitant Fuzzy Linguistic Preference Relation

Different types of preference relations have been addressed to solve the problems containing uncertain factors and people's vague thoughts [13, 18–20]. To overcome the limitations brought by the quantified preference relation, LPR was introduced [1, 2, 4]. As HFLPR [23] has advantages in describing the pairwise comparisons among several possible linguistic terms, it has been used for flexible information expressing in decision making problems.

Some researches proposed to obtain a ranking from a HFLPR through addressing the consistency definition, consistency checking, inconsistency repairing, and priorities deriving [7, 12, 14, 21–23]. Ren et al. [10] presented that directly obtaining the priority vector from a HFLPR is also significant for getting the decision making results succinctly and efficiently, especially in the problem with efficiency requirement. In the following, we show the method for deriving a ranking from a HFLPR introduced by Ren et al. [10].

6.1 Hesitant Fuzzy Linguistic Preference Relation

Definition 6.1 ([23]). Let $X = \{x_1, x_2, \ldots, x_n\}$ be a set of alternatives and $S = \{s_\alpha | \alpha = 0, 1, \ldots, 2\tau\}$ be a LTS, then a HFLPR is defined as $B = (b_{ij})_{n\times n} \subset X \times X$, where $b_{ij} = \left\{b_{ij}^s | s = 1, 2, \ldots, l_{b_{ij}}\right\}$ ($l_{b_{ij}}$ is the number of LTs in b_{ij}) is a HFLE, indicates the hesitant degrees to which x_i prefers to x_j. For all $i, j = 1, 2, \ldots, n$, b_{ij} $(i < j)$ should satisfy the conditions:

$$
\begin{gathered}
b_{ij}^{\sigma(s)} \oplus b_{ji}^{\sigma(s)} = s_{2\tau}, \quad b_{ii} = \{s_\tau\}, \quad l_{b_{ij}} = l_{b_{ji}}, \\
b_{ij}^{\sigma(s)} < b_{ij}^{\sigma(s+1)}, \quad b_{ji}^{\sigma(s+1)} < b_{ji}^{\sigma(s)}
\end{gathered}
\tag{6.1}
$$

P. Ren and Z. Xu, *Decision-Making Analyses with Thermodynamic Parameters and Hesitant Fuzzy Linguistic Preference Relations*, Studies in Fuzziness and Soft Computing 409, https://doi.org/10.1007/978-3-030-73253-0_6

where $b_{ij}^{\sigma(s)}$ and $b_{ji}^{\sigma(s)}$ are sth elements in b_{ij} and b_{ji}, respectively.

Motivated by the property of a consistency FPR $R = (r_{ij})_{n\times n} \subset X \times X$: $r_{ij} = \frac{w_i}{w_i+w_j}$ for $i, j = 1, 2, \ldots, n$, where $w = (w_1, w_2, \ldots, w_n)^T$ is the priority vector of R ([16, 17]), Ren et al. [10] proposed that a HFLPR $B = (b_{ij})_{n\times n} \subset X \times X$ is called to be consistent if

$$\frac{w_i}{w_i + w_j} = \frac{I(b_{ij})}{2\tau}, i, j = 1, 2, \ldots, n \tag{6.2}$$

where $I(b)$ is an element of the set $I = \left\{ I(b_{ij}^s) \middle| s = 1, 2, \ldots, l_{b_{ij}} \right\}$, and $w = (w_1, w_2, \ldots, w_n)^T$ is the priority vector derived from B satisfying $\sum_{i=1}^{n} w_i = 1$, $w_i \geq 0, i = 1, 2, \ldots, n$.

6.2 A Hyperplane-Consistency-Based Programming Model

6.2.1 Satisfaction for the Consistency Degree

Since a consistent HFLPR satisfies Eq. (6.2), Ren et al. [10] introduced a function as:

$$F_{ij}(w) = w_i - \frac{1}{2\tau}(w_i + w_j)I(b_{ij}) \tag{6.3}$$

Ren et al. [10] explained that the HFLPR is completely consistent if $F_{ij}(w) = 0$ for $i, j = 1, 2, \ldots, n$, and the HFLPR is called to be acceptably consistent if $F_{ij}(w) \approx 0$. Obviously, for the acceptably consistent HFLPR, we have

$$w_i - \frac{1}{2\tau}(w_i + w_j)I(b_{ij}) \approx 0 \tag{6.4}$$

The inconsistency threshold can be presented as:

Definition 6.2 ([10]). Let $B = (b_{ij})_{n\times n} \subset X \times X$ be a HFLPR, and $w = (w_1, w_2, \ldots, w_n)^T$ is the priority vector of B. Suppose that η is the inconsistency threshold, then B is called to be acceptably consistent if

$$\Phi(F_{ij}(w)) \leq \eta \tag{6.5}$$

where $\Phi(\cdot)$ is a function with the independent variable $F_{ij}(w)$.

Ren et al. [10] noted that the specific form of the function $\Phi(\cdot)$ is determined by decision makers. Some commonly used ways of the function $\Phi(\cdot)$ can be listed as: $\Phi(\cdot) = \max(\cdot)$, $\Phi(\cdot) = \frac{2}{n(n-1)} \sum_{i<j} |\cdot|$ for $i, j = 1, 2, \ldots, n$.

Considering that the satisfaction degree of a decision maker to the priority vector of a HFLPR can manifest the consistency degree of the HFLPR, and based on the idea of describing human intuitive cognition with FSs [8], Ren et al. [10] stated that if a HFLPR is consistent, then the satisfaction degree of a decision maker is 1, if a HFLPR is acceptably consistent, then the satisfaction degree of a decision maker decreases to some deviation limits. Ren et al. [10] further gave the following function to reflect the viewpoint:

$$S_{ij}(w) = \begin{cases} 1 - \frac{e^{F_{ij}(w)}-1}{t} & if\ F_{ij}(w) \geq 0 \\ 1 - \frac{e^{-F_{ij}(w)}-1}{t} & if\ F_{ij}(w) < 0 \end{cases} \tag{6.6}$$

The graph of the function can be drawn in Fig. 6.1.

Ren et al. [10] mentioned that since decision makers are more sensitive to the massive inconsistency than the slight inconsistency, the function of the satisfaction here is defined as a non-linear convex function with the increasing marginal utility. They furtherly described that

1. if $\frac{w_i}{w_i+w_j} = \frac{I(b_{ij})}{2\tau} \Leftrightarrow F_{ij}(w) = 0$, then $S_{ij}(w) = 1$, the decision maker perfectly satisfies with the priority vector;
2. if $0 < \left|\frac{w_i}{w_i+w_j} - \frac{I(b_{ij})}{2\tau}\right| \leq t \Leftrightarrow F_{ij}(w) \in [-t, t]$, then $0 \leq S_{ij}(w) < 1$, the decision maker imperfectly satisfies with the priority vector;

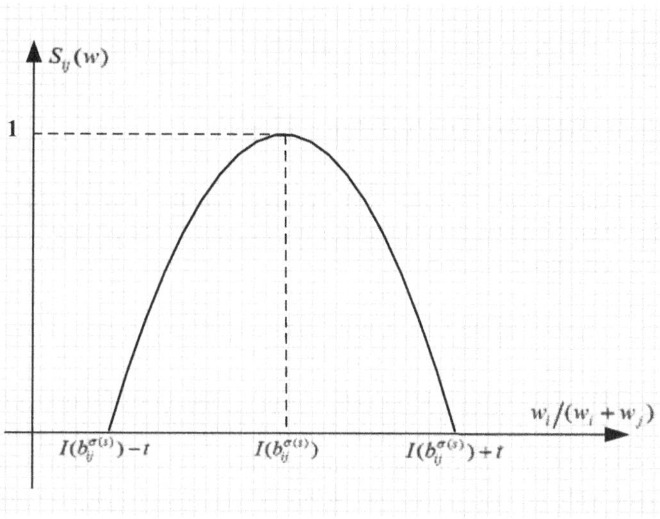

Fig. 6.1 The function of the satisfaction [10]

3. if $\left|\frac{w_i}{w_i+w_j} - \frac{I(b_{ij})}{2\tau}\right| > t \Leftrightarrow F_{ij}(w) \in (-\infty, -t)\&(t, \infty)$, then $S_{ij}(w) < 0$, the decision maker dissatisfies with the priority vector.

6.2.2 Priority Space

Ren et al. [10] proposed to apply the knowledge of hyperplane to analyze the priority space and defined a hyperplane in a n-dimensional priority space as:

$$P_{ij}(w) = \{w | P_{ij}(w) = 0\} \tag{6.7}$$

Then, a regular hyperplane in a $n-1$-dimensional priority space was introduced as [10]:

$$P^{n-1} = \{(w_1, w_2, \ldots, w_n) | w_1 + w_2 + \cdots + w_n = 1\} \tag{6.8}$$

Ren et al. [10] obtained a hyper line containing the priority information by intersecting the hyperplanes $P_{ij}(w)$ and P^{n-1}:

$$L_{ij}(w) = \min(P_{ij}(w), P^{n-1}) \tag{6.9}$$

To derive the solution of the prioritization problem, all hyper lines $L_{ij}(w)$ for all $i, j = 1, 2, \ldots, n$ are intersected [10]:

$$W(w) = \min\{L_{ij}(w) | i, j = 1, 2, \ldots, n, i < j\} \tag{6.10}$$

6.2.3 Mathematical Programming Model

By the relationship between the satisfaction degree and the priority vector shown in Eq. (6.6), it is easy to represent the satisfaction degrees to the priority vectors in the $n-1$-dimensional priority space [10]:

$$S_p(w) = \min_{w \in P^{n-1}} \{S_{ij}(w) | i, j = 1, 2, \ldots, n, \quad i < j\} \tag{6.11}$$

By Eq. (6.11), a series of priority vectors with a collection of satisfaction degrees can be obtained. Since there exists at least one solution of the convex function $S_{ij}(w)$, which corresponds to the maximum satisfaction degree [3], then there is at least one priority vector with the maximum consistency degree, which can be found by [10]

$$M_p(w) = \max \min_{w \in P^{n-1}} \{S_{ij}(w) | i, j = 1, 2, \ldots, n; i < j\} \tag{6.12}$$

Since $F_{ij}(w) = w_i - \frac{1}{2\tau}(w_i + w_j)I(b_{ij})$, then Eq. (4.12) can be transformed into the following mathematical programming model:

Model (Ren et al. [10]).

$$\max \lambda \begin{cases} s.t.\ t\lambda + e^{\left(w_i - \frac{1}{2\tau}(w_i+w_j)I(b_{ij})\right)} \leq t+1 \\ t\lambda + e^{-\left(w_i - \frac{1}{2\tau}(w_i+w_j)I(b_{ij})\right)} \leq t+1 \\ i, j = 1, \ldots, n, i < j \\ \sum_{i=1}^{n} w_i = 1, w_i \geq 0, i = 1, \ldots, n \end{cases} \tag{6.13}$$

Ren et al. [10] proved that Model 6.1 is solvable.

6.3 Decision-Making Procedure

For a MCDM problem with the alternatives A_i $(i = 1, 2, \ldots, m)$ and the criteria C_j $(j = 1, 2, \ldots, n)$, a decision maker compares n criteria in pair and provides a HFLPR, and compares m alternatives in pair with respect to n criteria and provides n HFLPRs. The decision making process for the problem can be briefly summarized [10]:

Step 1. Obtain the priority vector from the HFLPR on criteria by Model 6.1, and donate the priority vector of the criteria as $w_C = (w_{C_1}, w_{C_2}, \ldots, w_{C_n})^T$.

Step 2. Derive the priority vectors from the HFLPRs on alternatives by Model 6.1, and denote the priority vectors of alternatives regarding criteria as $w_A^{C_j} = \left(w_{A_1}^{C_j}, w_{A_2}^{C_j}, \ldots, w_{A_m}^{C_j}\right)^T$ for $j = 1, 2, \ldots, n$.

Step 3. Calculate the comprehensive priorities of alternatives by

$$w_{A_i} = \sum_{j=1}^{n} w_{C_j} \cdot w_{A_i}^{C_j} \text{ for } i = 1, 2, \ldots, m \tag{6.14}$$

and rank alternatives according to the comprehensive priorities. The best alternative corresponds to the maximum comprehensive priority.

The decision making process can be briefly shown in Fig. 6.2.

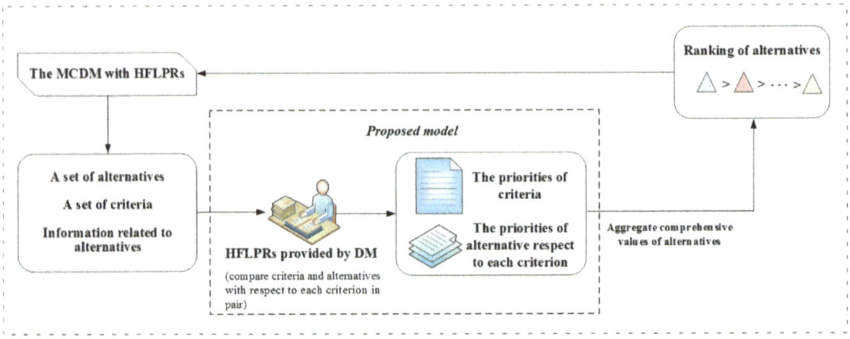

Fig. 6.2 The framework of the MCDM problem with HFLPRs [10]

6.4 Discussions

6.4.1 Analyses on Parameter t

As there is a parameter t in Model 6.1, Ren et al. [10] further made discussions on the value of t to make the model valid. Firstly, they transformed the constraint conditions of Model 6.1 as:

$$\begin{cases} t(1-\lambda)+1 \geq e^{\left(w_i - \frac{1}{2\tau}(w_i+w_j)I(b_{ij})\right)} \\ t(1-\lambda)+1 \geq e^{-\left(w_i - \frac{1}{2\tau}(w_i+w_j)I(b_{ij})\right)} \end{cases} \tag{6.15}$$

Since λ represents the minimum satisfaction degree of the decision maker, then $0 \leq \lambda \leq 1$. Since $t \geq 0$, then $t(1-\lambda)+1 \geq 1$, and Eq. (6.15) is equivalent to the following expressions [10]:

$$\begin{cases} \ln[t(1-\lambda)+1] \geq w_i - \frac{1}{2\tau}(w_i+w_j)I(b_{ij}) \\ \ln[t(1-\lambda)+1] \geq -\left(w_i - \frac{1}{2\tau}(w_i+w_j)I(b_{ij})\right) \end{cases} \tag{6.16}$$

Let $\xi = \ln[t(1-\lambda)+1]$, then $\lambda = 1 - \frac{e^{\xi}-1}{t}$, and Eq. (6.16) can be rewritten as [10]:

$$\min \xi$$
$$\begin{cases} s.t.\ \xi \geq w_i - \frac{1}{2\tau}(w_i+w_j)I(b_{ij}) \\ \quad \xi \geq -\left(w_i - \frac{1}{2\tau}(w_i+w_j)I(b_{ij})\right) \\ \quad i,j = 1,\ldots,n, i<j \\ \quad \sum_{i=1}^{n} w_i = 1, w_i \geq 0, i = 1,\ldots,n \end{cases} \tag{6.17}$$

Since $0 \leq \lambda \leq 1$, then $\lambda = 1 - \frac{e^{\xi}-1}{t} \geq 0$, which means $t \geq e^{\xi} - 1$, this is to say, $\min t = e^{\xi} - 1$. Based on these, Ren et al. [10] proposed to evaluate the value of $\min t$ through the value of ξ. To do so, they randomly constructed the 10,000 MCDM problems with HFLPRs, where there are five alternatives, and obtained the values of ξ in the 10,000 problems by Eq. (6.17), then $\min t$ can be got. Furthermore, they extended the same experimental process for the MCDM problems with three to ten alternatives, and obtained the densities of $\min t$ (as shown in Fig. 6.3).

In Fig. 6.3, different colors for the densities of $\min t$ correspond to the different orders of HFLPRs. By Fig. 6.3, the distribution of $\min t$ follows the normal distribution for each order of HFLPRs. Therefore, it is reasonable to use the arithmetic mean for the eigenvalue of $\min t$. Then, the suggested values of t in the programming model with different orders of HFLPRs can be listed in Table 6.1 [10].

By Fig. 6.3 and Table 6.1, the suggestible minimum value of t decreases while the order of the HFLPR increases, which indicates that the low-order HFLPR is more sensitive to the value of t.

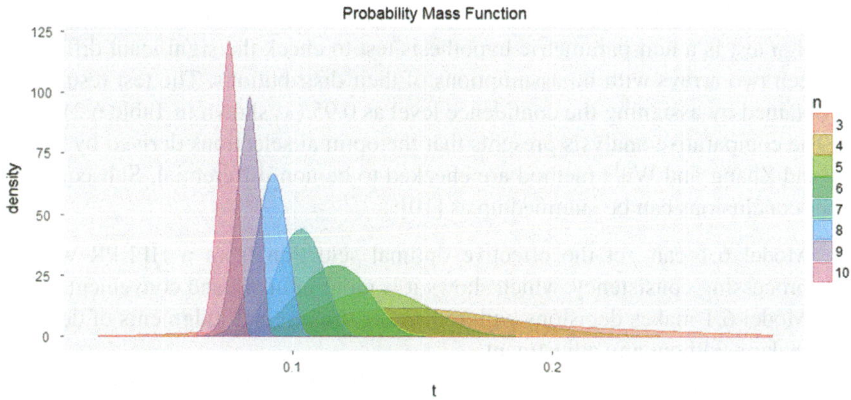

Fig. 6.3 The densities of $\min t$ for the 10,000 MCDM problems with HFLPRs' order varying from three to ten [10]

Table 6.1 The suggested value of t for each order of HFLPR [10]

The order of HFLPR	The minimum value of t	The order of HFLPR	The minimum value of t
3	0.1590	4	0.1463
5	0.1302	6	0.1147
7	0.1017	8	0.0909
9	0.0820	10	0.0746

6.4.2 Statistical Comparative Study

Considering that Model 6.1 directly derives the priorities without consistency checking, Ren et al. [10] verified the programming model's validation by comparing it with an existing method with consistency guarantees [22]. The idea of the comparison is to check the difference between the optimal selections derived by Model 6.1 and Zhang and Wu's method. If the optimal selections are non-differential, then Model 6.1 is considered to be valid.

Firstly, Ren et al. [10] put forward to acquire a series of optimal selections by the two methods through random simulation. More specifically, (1) randomly generate 1000 MCDM problems with HFLPRs, which consist of five alternatives, (2) apply Model 6.1 and Zhang and Wu's method to solve the 1000 problems, respectively; (3) collect a sequence of optimal selections of the 1000 problems by each method; (4) extend the above process to the decision making problems with three to ten alternatives.

Based on the collected optimal sequences of the two methods, Ren et al. [10] further utilized sign test [6] to check the difference between them, and explained that sign test is a non-parametric hypothesis test to check the significant difference between two arrays with no assumptions of their distributions. The test results can be obtained by assigning the confidence level as 0.95 (as shown in Table 6.2).

The comparative analysis presents that the optimal selections derived by Model 6.1 and Zhang and Wu's method are checked to be non-differential. Subsequently, other conclusions can be summed up as [10]:

1. Model 6.1 can get the effective optimal selection from a HFLPR without processing consistency, which shows it is more intuitive and convenient.
2. Model 6.1 makes decisions with remaining the original judgments of decision makers without any adjustment.
3. The time complexity of Model 6.1 is lower than Zhang and Wu's method, which indicates that Model 6.1 can achieve good application in practice.

Table 6.2 The sign test results between the optimal sequences derived by Model 6.1 and Zhang and Wu's method [10]

The order of HFLPRs	The test results	The order of HFLPRs	The test results	The significance conclusion
$n = 3$	Sig. 0.684	$n = 4$	Sig. 0.308	Sig. > 0.05 for $n = 3$ *to* 10. There exists no significant difference between Model 4.1 and the decision method with consistent HFLPRs
$n = 5$	Sig. 0.727	$n = 6$	Sig. 0.589	
$n = 7$	Sig. 0.366	$n = 8$	Sig. 0.213	
$n = 9$	Sig. 0.324	$n = 10$	Sig. 0.360	

6.5 Extension of the Model Under Incomplete Environment

6.5.1 Incomplete Hesitant Fuzzy Linguistic Preference Relation

Definition 6.3 ([11]). Let $X = \{x_1, x_2, \ldots, x_n\}$ be a set of alternatives and $S = \{s_\alpha | \alpha = 0, 1, \ldots, 2\tau\}$ (τ is a positive number) be a LTS, then an incomplete HFLPR (IHFLPR) B on S is presented by a matrix $B = (b_{ij})_{n\times n} \subset X \times X$, in which all known HFLEs $b_{ij} = \left\{b_{ij}^s | s = 1, 2, \ldots, l_{b_{ij}}\right\}$ ($l_{b_{ij}}$ is the number of linguistic terms in b_{ij}) indicate the possible preference degrees of x_i over x_j and satisfy the condition in a HFLPR.

It is noted that the calculations of elements in IHFLPRs follow the operational rules of HFLTSs.

6.5.2 A Programming Model for Deriving Priorities

The section extends the programming model in Sect. 6.2 into the MCDM problems with IHFLPRs. Firstly, the additive consistency and multiplicative consistency of an IHFLPR can be presented as:

Definition. 6.4 [9]. Let $B = (b_{ij})_{n\times n}$ be an IHFLPR, where $b_{ij} = \left\{b_{ij}^s | s = 1, 2, \ldots, l_{b_{ij}}\right\}$, then B is called to be additively consistent if all known HFLEs satisfy

$$\frac{1}{2\tau}\left(I\left(b_{ij}^{\sigma(1)}\right) \text{ or } \ldots \text{ or } I\left(b_{ij}^{\sigma(l_{b_{ij}})}\right)\right) = 0.5\left(w_i - w_j + 1\right), i, j \in \{1, 2, \ldots, n\} \tag{6.18}$$

Definition 6.5 [9]. Let $B = (b_{ij})_{n\times n}$ be an IHFLPR, where $b_{ij} = \left\{b_{ij}^s | s = 1, 2, \ldots, l_{b_{ij}}\right\}$, then B is called to be multiplicatively consistent if all known HFLEs satisfy.

$$\frac{1}{2\tau}\left(I\left(b_{ij}^{\sigma(1)}\right) \text{ or } \ldots \text{ or } I\left(b_{ij}^{\sigma(l_{b_{ij}})}\right)\right) = \frac{w_i}{w_i + w_j}, i, j = 1, 2, \ldots, n \tag{6.19}$$

Denoting $\psi(b_{ij}) = I\left(b_{ij}^{\sigma(1)}\right)$ or $\ldots$ or $I\left(b_{ij}^{\sigma(l_{b_{ij}})}\right)$, then $\delta_{ij}\left(\frac{1}{2\tau}\psi(b_{ij}) - 0.5\left(w_i - w_j + 1\right)\right)$ approaching to 0 indicates the IHFLPR achieve high consistency [9], where $\Delta = (\delta_{ij})_{n\times n}$ is an indication matrix, and $\delta_{ij} =$

$\begin{cases} 0, & b_{ij} \text{ is a missing HFLE} \\ 1, & b_{ij} \text{ is not a missing HFLE} \end{cases}$ ([15]). Based on the additive consistency, the satisfaction degree of a decision maker to the priority vector of an IHFLPR can be introduced as [9]:

$$SD_{ij}(w) = \begin{cases} 1 - \frac{\delta_{ij}\left(\frac{1}{2\tau}\psi(b_{ij}) - 0.5\,(w_i - w_j + 1)\right)}{d_{ij}}, & if \frac{1}{2\tau}\psi\left(b_{ij}\right) > 0.5\left(w_i - w_j + 1\right) \\ 1 + \frac{\delta_{ij}\left(\frac{1}{2\tau}\psi(b_{ij}) - 0.5\,(w_i - w_j + 1)\right)}{d_{ij}}, & if \frac{1}{2\tau}\psi\left(b_{ij}\right) < 0.5\left(w_i - w_j + 1\right) \end{cases} \tag{6.20}$$

where d_{ij} ($d_{ij} > 0$) is the deviation variable of b_{ij}.

Similarly, a regular $n-1$-simplex can be used to present the priority space [9]:

$$SX^{n-1} = \{\,(w_1, w_2, \ldots, w_n)|w_1 + w_2 + \cdots + w_n = 1\,\} \tag{6.21}$$

The hyper line with priority information can be obtained by the intersection of the hyperplanes, which can be transferred into the following representation based on the expression of satisfaction degree [9]:

$$SD = \min_{w \in SX^{n-1}} \left\{SD_{ij}(w)|i, j = 1, 2, \ldots, n\right\} \tag{6.22}$$

The optimal priority vector corresponds to the maximum satisfaction degree, which is [9]

$$\max_{i,j=1,\ldots,n} \min_{w \in SX^{n-1}} \left\{SD_{ij}(w)|i, j = 1, 2, \ldots, n\right\} \tag{6.23}$$

A mathematical programming model can be derived based on Eq. (6.23):

Model 6.2 (Ren et al. [9]).

$$\begin{aligned} & \max \gamma \\ s.t.\ & d\gamma - 0.5\,\delta_{ij} w_i + 0.5\,\delta_{ij} w_j \le d + 0.5\,\delta_{ij} \\ & - \frac{\delta_{ij}}{2\tau}\left(I\left(b_{ij}^{\sigma(1)}\right)\ or\ \ldots\ or\ I\left(b_{ij}^{\sigma(l_{b_{ij}})}\right)\right) \\ & d\gamma + 0.5\,\delta_{ij} w_i - 0.5\,\delta_{ij} w_j \le d - 0.5\,\delta_{ij} \\ & + \frac{\delta_{ij}}{2\tau}\left(I\left(b_{ij}^{\sigma(1)}\right)\ or\ \ldots\ldots\ or\ I\left(b_{ij}^{\sigma(l_{b_{ij}})}\right)\right) \\ & i, j = 1, 2, \ldots, n;\ i \ne j \\ & \sum_{i=1}^{n} w_i = 1,\ w_i \ge 0. \end{aligned} \tag{6.24}$$

In the same way, the satisfactory function of a decision maker in regard to the priority vector of an IHFLPR based on its multiplicative consistency can be given as [9]:

$$SD_{ij}(w)=\begin{cases}1-\dfrac{\delta_{ij}((\frac{1}{2\tau}\psi(b_{ij})-1)w_i+\frac{1}{2\tau}\delta_{ij}\psi(b_{ij})w_j)}{v_{ij}}, \\ \quad if\left(\dfrac{1}{2\tau}\psi(b_{ij})-1\right)w_i+\dfrac{1}{2\tau}\delta_{ij}\psi(b_{ij})w_j>0 \\ 1+\dfrac{\delta_{ij}((\frac{1}{2\tau}\psi(b_{ij})-1)w_i+\frac{1}{2\tau}\delta_{ij}\psi(b_{ij})w_j)}{v_{ij}}, \\ \quad if\left(\dfrac{1}{2\tau}\psi(b_{ij})-1\right)w_i+\dfrac{1}{2\tau}\delta_{ij}\psi(b_{ij})w_j<0\end{cases} \tag{6.25}$$

where v_{ij} $(v_{ij}>0)$ is the deviation variable of b_{ij}.

By defining the priority space, a mathematical programming model for deriving the optimal priority vector from an IHFLPR based on its multiplicative consistency can be established as:

Model 6.3 (Ren et al. [9]).

$$\begin{aligned}&\max \xi\\ s.t.\ & v\xi+\delta_{ij}\left(\frac{1}{2\tau}\left(I\left(b_{ij}^{\sigma(1)}\right)\ or\ \ldots\ or\ I\left(b_{ij}^{\sigma(l_{b_{ij}})}\right)\right)-1\right)w_i\\ &+\frac{1}{2\tau}\delta_{ij}(I(b_{ij}^{\sigma(1)})\ or\ \ldots or\ I(b_{ij}^{\sigma(l_{b_{ij}})}))w_j\leq v\\ &v\xi-\delta_{ij}(\frac{1}{2\tau}(I(b_{ij}^{\sigma(1)})\ or\ \ldots or\ I(b_{ij}^{\sigma(l_{b_{ij}})}))-1)w_i\\ &-\frac{1}{2\tau}\delta_{ij}(I(b_{ij}^{\sigma(1)})\ or\ \ldots or\ I(b_{ij}^{\sigma(l_{b_{ij}})}))w_j\leq v\\ &i,j=1,2,\ldots,n;\ \ i\neq j\\ &\sum_{i=1}^{n}w_i=1,\ w_i\geq 0.\end{aligned} \tag{6.26}$$

Both Models 6.2 and 6.3 can be applied in the decision making procedure in Sect. 6.3.

6.5.3 Discussions

The section aims to present the comparative analyses of the two proposed models with IHFLPRs based on different consistency properties.

Ren et al. [9] gave the steps for obtaining the fundamental comparative data:

Step 1. Randomly produce 1000 IHFLPRs with the order $n = 3$;

Step 2. Obtain the rankings from the produced IHFLPRs by Models 4.2 and 4.3, and record the sequences of the optimal selections derived from the two models, respectively;

Step 3. Let n vary from 4 to 10, repeat Step 1 and Step 2.

Subsequently, Ren et al. [9] provided the following perspectives to make comparisons:

Perspective* 1. *Complexity

Ren et al. [9] proposed the average time for processing the produced 1000 IHFLPRs to derive the priority vectors as complexity. If the average time proceeded by Model 6.2 is less than Model 6.3, then Model 6.2 is considered to be more efficient. The average time for processing different orders of IHFLPRs by Matlab can be shown in Fig. 6.4.

In Fig. 6.4, the ordinate records the average time for the produced 1000 IHFLPRs. It is evident that Model 6.3 is superior to Model 6.2 in simplicity and convenience since the complexity test of Model 6.3 is less than Model 6.2 [9].

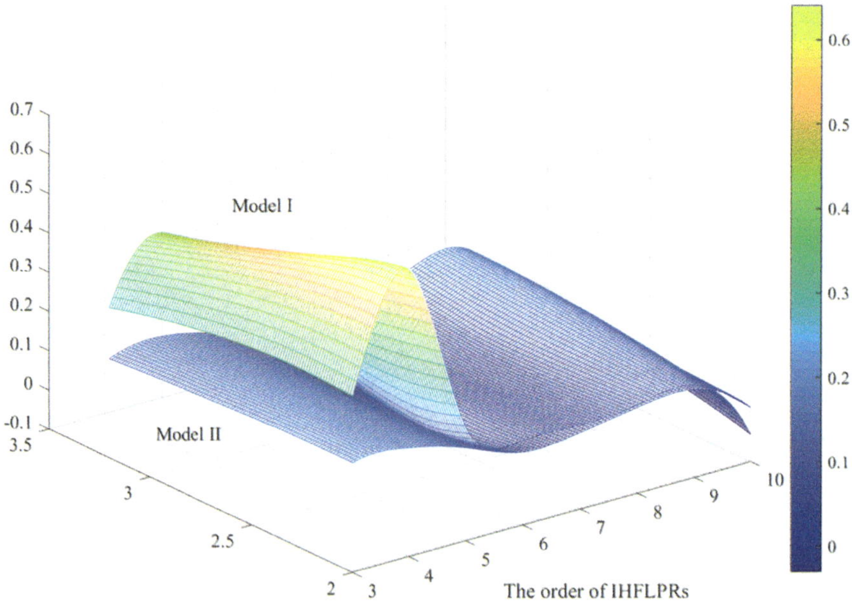

Fig. 6.4 Complexity test of Models 6.2 and 6.3 [9]

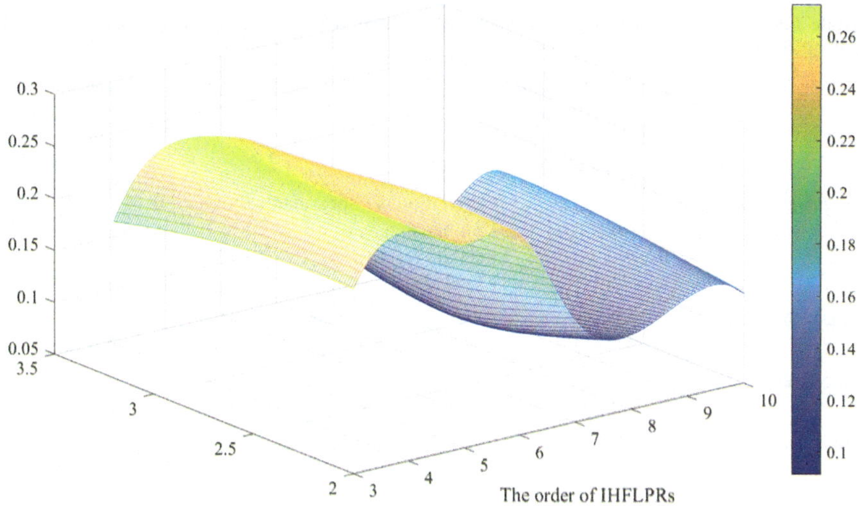

Fig. 6.5 The ratio of the same optimal selections derived by Models 6.2 and 6.3 [9]

Perspective 2. Different solutions

This part directly discusses the difference between the optimal sequences derived from Models 6.2 and 6.3. Firstly, Ren et al. [9] presented the ratio of the same optimal selections by the two models with each order of IHFLPR in Fig. 6.5.

In Fig. 6.5, the ordinate records the ratio of the same optimal selections by Models 6.2 and 6.3. For $n = 3, \ldots, 10$, the ratios in Fig. 6.5 are 22.90%, 27.20%, 24.30%, 24.50%, 11.10%, 9.60%, 13.50% and 16.10%, respectively. Furtherly, Ren et al. [9] put forward to apply Spearman correlation coefficient [5] to analyze the correlation between the optimal sequences derived from the two models, and listed the results in Table 6.3.

By Table 6.3, there is no correlation between the optimal sequences obtained by the two models with Spearman correlation analysis because they failed the test of significance generally. It can be accessed that despite the same modeling idea is utilized in Models 6.2 and 6.3, the obtained optimal sequences are independent with different consistency properties, and the independence is not affected by the orders of IHFLPR, which can be called "*Consistency Dominance Effect*" [9].

Table 6.3 Spearman correlation coefficient of optimal selection sequences for each order of IHFLPRs [9]

	Correlation coefficient Sig. (2-tailed)			Correlation coefficient Sig. (2-tailed)
Spearman's rho ($n = 3$)	−0.165 0.000	Correlation is significant at the 0.05 level (2-tailed)	Spearman's rho ($n = 4$)	0.008 0.807
Spearman's rho ($n = 5$)	0.046 0.143		Spearman's rho ($n = 6$)	0.059 0.061
Spearman's rho ($n = 7$)	−0.086 0.061		Spearman's rho ($n = 8$)	0.024 0.451
Spearman's rho ($n = 9$)	0.034 0.286		Spearman's rho ($n = 10$)	0.000 0.980

6.6 A Case Study on Assessing the Effects of Hydropower Stations' Flood Discharge and Energy Dissipation on Environment

6.6.1 Description of the Case

Model 6.1 for HFLPRs was applied to assess the effects of hydropower stations' flood discharge and energy dissipation on the environment. The background was introduced as [10]:

Hydropower station is constructed to transform hydro energy into electric energy, which also performs retaining water, releasing water, diverting water, etc. The benefits brought by it can be briefly presented as (1) flood control benefit (Regulating water storage and improving flood control standard); (2) electric power provided benefit (solving the power consumption of residents); (3) river navigation improvement benefit (improving waterway regulation); (4) cultural benefit; among others.

It should be noted that the construction of hydropower stations still brings adverse effects such as influencing the ecological environment, resettling inhabitants. Thereinto, the flood discharge and energy dissipation, as a process during the operation of the hydropower station, brings some adverse effects on the environment, including

1. Shaking. The released energy transmits through the ground, damages the surrounding buildings, and reduces residents' living quality.
2. Atomization. The flood discharge and energy dissipation increase atmospheric moisture.
3. Erosion. The flood discharge and energy dissipation impact the downstream river channel.

Table 6.4 A HFLPR between any two negative aspects [10]

	Shaking	Atomization	Erosion	Aeration
Shaking	$\{s_4\}$	$\{s_2, s_3\}$	$\{s_2, s_3, s_4\}$	$\{s_3, s_4\}$
Atomization		$\{s_4\}$	$\{s_5, s_6\}$	$\{s_5, s_6\}$
Erosion			$\{s_4\}$	$\{s_2, s_3\}$
Aeration				$\{s_4\}$

4. Aeration. The flood discharge and energy dissipation infuse excessive gas into water.

As Sichuan China possesses many hydropower stations, it is necessary to assess the flood discharge and energy dissipation's impact on the environment to reduce environmental change. To do so, Ren et al. [10] proposed to use Model 6.1 for HFLPRs to address the assessment of the hydropower stations in Sichuan. The hydropower stations were selected as Xiangjiaba hydropower station, Xiluodu hydropower station, Pubugou hydropower station, and Ertan hydropower station. Suppose that an expert panel is invited to make pairwise comparisons between any two negative aspects and any two hydropower stations, and provides the HFLPRs as:

Where all HFLPRs are given based on the LTS:

$$S = \left\{ \begin{array}{lll} s_0\text{: extemely poor}, & s_1\text{: very poor}, & s_2\text{: poor} \\ s_3\text{: slightly poor}, & s_4\text{: fair}, & s_5\text{: slightly good} \\ s_6\text{: good}, & s_7\text{: very good}, & s_8\text{: extremely good} \end{array} \right\}.$$

6.6.2 Decision-Making Process

Ren et al. [10] presented the following procedure for the above decision making problem:

Step 1. Apply Model 6.1 to obtain the priority vector of the four negative aspects based on Table 6.4:

$$\begin{aligned} \boldsymbol{w} &= (\text{Shaking Atomization Erosion Aeration})^T \\ &= (0.1628, 0.4419, 0.1628, 0.2325)^T \end{aligned} \tag{6.27}$$

Step 2. Apply Model 6.1 to obtain the priority vector of the four hydropower stations with respect to each negative aspect based on Table 6.5. The results are shown in Table 6.6.

Step 3. Apply Eq. (6.14) to obtain the overall priority vector of the four hydropower stations:

$$\boldsymbol{w}^O = (\text{Xiangjiaba, Xiluodu, Pubugou, Ertan})^T$$

Table. 6.5 HFLPRs between any two hydropower stations with respect to negative effects [10]

		Xiangjiaba	Xiluodu	Pubugou	Ertan
Shaking	Xiangjiaba	$\{s_4\}$	$\{s_5, s_6, s_7\}$	$\{s_5, s_6\}$	$\{s_5, s_6\}$
	Xiluodu		$\{s_4\}$	$\{s_3, s_4, s_5\}$	$\{s_3, s_4\}$
	Pubugou			$\{s_4\}$	$\{s_4, s_5\}$
	Ertan				$\{s_4\}$
Shaking	Xiangjiaba	$\{s_4\}$	$\{s_2, s_3\}$	$\{s_5, s_6\}$	$\{s_4, s_5\}$
	Xiluodu		$\{s_4\}$	$\{s_5, s_6\}$	$\{s_5, s_6\}$
	Pubugou			$\{s_4\}$	$\{s_2, s_3, s_4\}$
	Ertan				$\{s_4\}$
Erosion	Xiangjiaba	$\{s_4\}$	$\{s_4, s_5\}$	$\{s_2, s_3\}$	$\{s_3, s_4\}$
	Xiluodu		$\{s_4\}$	$\{s_2, s_3\}$	$\{s_3, s_4\}$
	Pubugou			$\{s_4\}$	$\{s_2\}$
	Ertan				$\{s_4\}$
Aeration	Xiangjiaba	$\{s_4\}$	$\{s_4, s_5\}$	$\{s_4, s_5\}$	$\{s_2, s_3\}$
	Xiluodu		$\{s_4\}$	$\{s_4, s_5\}$	$\{s_2, s_3, s_4\}$
	Pubugou			$\{s_4\}$	$\{s_3, s_4\}$
	Ertan				$\{s_4\}$

Table. 6.6 The priority vector of the four hydropower stations with respect to each negative aspect [10]

	Xiangjiaba	Xiluodu	Pubugou	Ertan
Shaking	0.4261	0.1420	0.1989	0.2330
Atomization	0.2174	0.4130	0.1522	0.2174
Erosion	0.1957	0.1957	0.2391	0.3696
Aeration	0.2083	0.2083	0.2361	0.3472

$$= (0.2457, 0.2859, 0.1935, 0.2749)^T \tag{6.28}$$

The ranking of the hydropower stations is Xiluodu $\succ$ Ertan $\succ$ Xiangjiaba $\succ$ Pubugou. The results indicate that Xiluodu hydropower station's flood discharge and energy dissipation have the lightest adverse effects on the environment. The flood discharge and energy dissipation project of Pubugou hydropower station should be paid more attention to be improved, especially on atomization [10].

6.7 Summary

Considering the feasibility of the HFLPRs in describing the judgments through pairwise comparison way and people's expression habit with the fuzzy mind, the chapter has demonstrated a priority programming method for HFLPRs [10], which contributes to:

1. Address the decision scenario that naturally expresses judgments by the obscure LTs under uncertainty, and reflect the more sensitivity of people towards the lower satisfaction degrees;
2. Directly and efficiently derive the priority vector from a HFLPR without consistency checking and inconsistency improving;
3. Design experiments to analyze the parameter t in the proposed programming model under different decision making environments, which provides decision support for the proposed model;
4. Extend the proposed model into an incomplete environment to improve the applicability.

References

1. Dong, Y.C., Hong, W.C., Xu, Y.: Measuring consistency of linguistic preference relations: a 2-tuple linguistic approach. Soft. Comput. **17**, 2117–2130 (2013)
2. Dong, Y.C., Xu, Y.F., Li, H.Y.: On consistency measures of linguistic preference relations. Eur. J. Oper. Res. **189**(2), 430–444 (2008)
3. Dubois, D., Fortemps, P.: Computing improved optimal solutions to max-min flexible constraint satisfaction problems. Eur. J. Oper. Res. **239**, 794–801 (1999)
4. Herrera, F., Herrera-Viedma, E., Verdegay, J.L.: A sequential selection process in group decision making with a linguistic assessment approach. Inf. Sci. **85**(4), 223–239 (1995)
5. Lehman, A., O'Rourke, N., Hatcher, L., Stepanski, E.: JMP for Basic Univariate and Multivariate Statistics: Methods for Researchers and Social Scientists. Cary, NCUnited States, SAS Institute Inc., SAS Campus Dr (2013)
6. Lehmann, E.: Nonparametrics: Statistical Methods Based on Ranks. Holden-Day, San Francisco (1975)
7. Mi, X.M., Wu, X.L., Tang, M., Liao, H.C., Al-Barakati, Altalhi, A.H., Herrera, F.: Hesitant fuzzy linguistic analytic hierarchical process with prioritization, consistency checking, and inconsistency repairing. IEEE Access **7**, 44135–44149 (2019)
8. Mikhailov, L.: Fuzzy analytical approach to partnership selection in formation of virtual enterprises. Omega **30**, 393–401 (2002)
9. Ren, P.J., Hao, Z.N., Wang, X.X., Zeng, X.-J., Xu, Z.S.: decision making models based on incomplete hesitant fuzzy linguistic preference relation with application to site selection of hydropower stations. IEEE Trans. Eng. Manage. (2020). https://doi.org/10.1109/TEM.2019.2962180
10. Ren, P.J., Zhu, B., Xu, Z.S.: Assessment of the impact of hydropower stations on the environment with a hesitant fuzzy linguistic hyperplane-consistency programming method. IEEE Trans. Fuzzy Syst. **26**(5), 2981–2992 (2018)
11. Song, Y.M., Li, G.X.: A mathematical programming approach to manage group decision making with incomplete hesitant fuzzy linguistic preference relations. Comput. Ind. Eng. **135**, 467–475 (2019)

12. Wu, P., Zhu, J.M., Zhou, L.G.: Automatic iterative algorithm with local revised strategies to improve the consistency of hesitant fuzzy linguistic preference relations. Int. J. Fuzzy Syst. **21**(7), 2283–2298 (2019)
13. Xia, M.M., Xu, Z.S.: Managing hesitant information in GDM problems under fuzzy and multiplicative preference relation. Internat. J. Uncertain. Fuzziness Knowledge-Based Systems **21**(6), 865–897 (2013)
14. Xu, Y., Wen, X., Sun, H., Wang, H.M.: Consistency and consensus models with local adjustment strategy for hesitant fuzzy linguistic preference relations. Int. J. Fuzzy Syst. **20**(7), 2216–2233 (2018)
15. Xu, Z.S.: Goal programming models for obtaining the weight vector of incomplete fuzzy preference relation. Int. J. Approximate Reasoning **36**, 261–270 (2004)
16. Xu, Z.S.: Uncertain Multiple Attribute Decision Making: Methods and Applications. Tsinghua University Press, Beijing (2004)
17. Xu, Z.S.: Intuitionistic preference relations and their application in group decision making. Inf. Sci. **177**, 2363–2379 (2007)
18. Xu, Z.S.: A survey of preference relations. Int. J. Gen Syst **36**, 179–203 (2007)
19. Xu, Z.S.: Consistency of interval fuzzy preference relations in group decision making. Appl. Soft Comput. **11**(5), 3898–3909 (2011)
20. Xu, Z.S., Chen, J.: Some models for deriving the priority weights from interval fuzzy preference relations. Eur. J. Oper. Res. **184**(1), 266–280 (2008)
21. Zhang, Z.M., Chen, S.M.: Group decision making based on acceptable multiplicative consistency and consensus of hesitant fuzzy linguistic preference relations. Inf. Sci. **541**, 531–550 (2020)
22. Zhang, Z.M., Wu, C.: On the use of multiplicative consistency in hesitant fuzzy linguistic preference relations. Knowl.-Based Syst. **72**, 13–27 (2014)
23. Zhu, B., Xu, Z.S.: Consistency measures for hesitant fuzzy linguistic preference relations. IEEE Transaction on Fuzzy Systems **22**, 34–45 (2014)

Chapter 7
A Group Decision-Making Method for Hesitant Fuzzy Linguistic Preference Relations Based on Modified Extent Measurement

Since group intelligence contributes to the effectiveness of the Decision making process, group decision making is important for the problems with uncertainties. Considering that different decision makers probably have different knowledge backgrounds and cognitions of things, dividing decision makers into different groups, obtaining results in each group, and integrating them into an overall one helps guarantee the scientificity of the results.

It is noted that although the clustering of decision makers can improve professionalism in each group, it is probably that different decision makers hold different opinions. Then, the compatibility of judgments in each clustering of decision makers is necessary to be measured to acquire reliable Decision making results.

Given that decision makers are limited by their thinking and knowledge, they may not be able to fully understand and agree with others. In such a case, letting decision makers always approach the overall judgments sometimes may exceed their cognition. As HFLPR was proposed as one of the efficient tools for portraying the preference of decision makers [13], the chapter will introduce a group Decision making method with HFLPR investigated by Ren et al. [7], which establishes group consensus based on measuring the modified extent of individual judgments.

7.1 A Kernel-Based Algorithm Under Hesitant Fuzzy Linguistic Environment

The section provides a kernel-based algorithm for clustering hesitant fuzzy linguistic information built by Ren et al. [7].

Suppose that h HFLPRs $B_1, B_2, \ldots, B_h$ are needed to be divided into c $(1 < c < n)$ clusters, by defining a nonlinear map as: Φ: $x \rightarrow \Phi(x) \in F$, where F is a feature space with a higher dimension, then a target function for clustering hesitant fuzzy linguistic information is:

P. Ren and Z. Xu, *Decision-Making Analyses with Thermodynamic Parameters and Hesitant Fuzzy Linguistic Preference Relations*, Studies in Fuzziness and Soft Computing 409, https://doi.org/10.1007/978-3-030-73253-0_7

$$J_{KHFLC}(U, V) = \sum_{q=1}^{c}\sum_{k=1}^{h} u_{qk}^{p}\left|\Phi(B_k) - \Phi(v_q)\right|^2 \tag{7.1}$$

subject to

$$\begin{cases} 0 \le u_{qk} \le 1, \quad 1 \le k \le h, \quad 1 \le q \le c \\ \sum\limits_{q=1}^{c} u_{qk} = 1, \quad \forall k \end{cases}$$

where $v_1, v_2, \ldots, v_c$ are the centroids of the c clusters, p $(1 \le p \le \infty)$ is a fuzzification parameter, and u_{qk} presents the membership degree that the HFLPR B_k belongs to the cluster v_q [7].

Since a kernel function satisfies $K(x, x) = 1$, then in Eq. (7.1),

$$\begin{aligned} \left|\Phi(B_k) - \Phi(v_q)\right|^2 &= \Phi(B_k)\Phi(B_k) - 2\Phi(B_k)\Phi(v_q) + \Phi(v_q)\Phi(v_q) \\ &= K(B_k, B_k) - 2K(B_k, v_q) + K(v_q, v_q) = 2(1 - K(B_k, v_q)) \end{aligned} \tag{7.2}$$

Ren et al. [7] proposed to choose the Gaussian kernel function [3] as a kernel function and expressed Eq. (7.1) as:

$$J_{KHFLC}(U, V) = 2\sum_{q=1}^{c}\sum_{k=1}^{h} u_{qk}^{p}\left(1 - \exp\left(\frac{-d^2(B_k, v_q)}{2\gamma^2}\right)\right) \tag{7.3}$$

subject to

$$\begin{cases} 0 \le u_{qk} \le 1, \quad 1 \le k \le h, \quad 1 \le q \le c \\ \sum\limits_{q=1}^{c} u_{qk} = 1, \quad \forall k \end{cases}$$

where $d^2(B_k, v_q)$ is the Euclidean distance between B_k and v_q, and γ is a free parameter.

It is noted that based on distance measures for HFLTSs [4, 13], Ren et al. [7] gave the Euclidean distance of any two hesitant fuzzy linguistic matrices (HFLMs) $B_1 = (b_{1,ij})_{m\times n}$ and $B_2 = (b_{2,ij})_{m\times n}$, in which $b_{1,ij} = \left\{b_{1,ij}^{s} | s = 1, 2, \ldots, l_b\right\}$ and $b_{2,ij} = \left\{b_{2,ij}^{s} | s = 1, 2, \ldots, l_b\right\}$ are HFLTSs with the same length as:

$$d(B_1, B_2) = \frac{1}{l_b \times m \times n}\sum_{i=1}^{m}\sum_{j=1}^{n}\sum_{s=1}^{l_b} \frac{\left|I^{\sigma(s)}(b_{1,ij}) - I^{\sigma(s)}(b_{2,ij})\right|^{\chi}}{2\tau + 1} \tag{7.4}$$

Based on the above preparations, the Kernel-based algorithm for clustering hesitant fuzzy linguistic information can be listed:

Algorithm 7.1 [7]. **Kernel-based hesitant fuzzy linguistic clustering.**

Input: HFLPRs $B_1, B_2, \ldots, B_h$, fuzzification parameter p, free parameter γ, a constant value ξ.

Output: Membership degrees of B_k $(k = 1, 2, \ldots, h)$ belongs to v_q $(q = 1, 2, \ldots, c)$.

Step 1. Initialize the centroids clusters $v_q(0)$ for $q = 1, 2, \ldots, c$;

Step 2. Let $t = 1$, and compute the membership degrees $u_{qk}(t)$ by

$$u_{qk}(t) = \frac{(1/1 - K(B_k, v_q(t)))^{1/p-1}}{\sum_{q=1}^{c}(1/1 - K(B_k, v_q(t)))^{1/p-1}} \tag{7.5}$$

for $k = 1, 2, \ldots, h$ and $q = 1, 2, \ldots, c$;

Step 3. Update the clustering centroids $v_q(t)$ for $q = 1, 2, \ldots, c$ by

$$v_q(t) = \frac{\sum_{k=1}^{h} u_{qk}^{p} K(B_k, v_q(t)) B_k}{\sum_{k=1}^{h} u_{qk}^{p} K(B_k, v_q(t))} \tag{7.6}$$

Step 4. If $V(t) - V(t-1) > \xi$, where $V(t) = [v_1(t), v_2(t), \ldots, v_c(t)]$ and $V(t-1) = [v_1(t-1), v_2(t-1), \ldots, v_c(t-1)]$, then let $t = t + 1$, return Step 2; otherwise, end.

7.2 Group Consensus Measurement

Since that letting individual judgments always approaching the overall judgments during the consensus process may exceed the cognitions of decision makers in some situations, as investigated in [1, 2, 5, 6, 8–12], Ren et al. [7] proposed that measuring the modified extent of individual judgments is a way to improve the weakness.

Definition 7.1 [7]. Suppose that during the consensus process, a decision maker's HFLPR gradually approaching the overall HFLPR, and the individual HFLPRs at the τth and the $\tau + 1$th rounds $(\tau \geq 0)$ are $B(\tau)$ and $B(\tau + 1)$, respectively, then the modified extent between the τth and the $\tau + 1$th rounds is.

$$e(\tau) = d(B(\tau), B(\tau + 1)) \tag{7.7}$$

where $d(B(\tau), B(\tau + 1))$ is the distance measure between the HFLPRs $B(\tau)$ and $B(\tau + 1)$.

Definition 7.2 [7]. Let $B(\tau)$ and $B(\tau+1)$ be the HFLPRs at the τth and the $\tau+1$th rounds ($\tau \geq 0$), and ρ be the maximum modified extent, then $B(\tau)$ is with acceptable consensus if.

$$e(\tau) \leq \rho \tag{7.8}$$

For obtaining the HFLPR in a new round, Ren et al. [7] proposed that if $B(\tau)$ is an individual HFLPR at the τth rounds and $\overline{B}$ is the overall HFLPR, then the $\tau+1$th round of individual HFLPR can be obtained by

$$B(\tau+1) = f(d(\tau))B(\tau) + (1 - f(d(\tau)))\overline{B} \tag{7.9}$$

where $d(\tau)$ is the distance measure between $B(\tau)$ and $\overline{B}$, $f(d(\tau))$ is a decreasing function which satisfies $f(0) = 1$, $f(1) = 0$, $f(d(\tau)) \in [0, 1][0, 1]$.

$f(d(\tau))$ is a function that is determined by decision makers according to their attitudes towards the distance between $B(\tau)$ and $\overline{B}$. If decision makers are sensitive to the distance, then $f(d(\tau))$ would be a convex function; or else $f(d(\tau))$ would be a concave function [7].

Based on the above consensus analysis, the following algorithm for the modified extent-based group decision making can be given:

Algorithm 7.2 [7]. **Modified extent-based group decision making with HFLPRs.**

Input: The individual HFLPRs B_k $(k = 1, 2, \ldots, h)$ of decision makers D_k $(k = 1, 2, \ldots, h)$, the weights of decision makers $\eta = (\eta_1, \eta_2, \ldots, \eta_h)^T$, the maximum modified extent ρ, and the parameter function $f(d(\tau))$.

Output: Each individual HFLPR with acceptable consensus.

Step 1. Compute the overall HFLPR $\overline{B}$ of all HFLPRs B_k for $k = 1, 2, \ldots, h$ by an aggregation operator;

Step 2. Initialize $\tau = 0$, denote B_k as $B_k(\tau)$ for $k = 1, 2, \ldots, h$;

Step 3. Calculate the distances between individual HFLPR $B_k(\tau)$ and $\overline{B}$, denoted as $d_k(\tau)$ for $k = 1, 2, \ldots, h$;

Step 4. Calculate $B_k(\tau+1)$ by Eq. (7.9) for $k = 1, 2, \ldots, h$,

Step 5. Obtain the individual modified extents for the HFLPRs at τth and $\tau+1$th rounds by Eq. (7.8), denoted as $e_k(\tau)$,

Step 6. Leave the HFLPRs with satisfying $e_k(\tau) \leq \rho$ unchanged, and output them at τth round. For the HFLPRs with satisfying $e_k(\tau) > \rho$, let $\tau = \tau + 1$, return Step 3.

7.3 A Group Decision-Making Procedure

Suppose that a group Decision making problem contains the alternatives A_i $(i = 1, 2, \ldots, n)$, the criteria C_x $(x = 1, 2, \ldots, m)$ with the weights $w =$

$(w_1, w_2, \ldots, w_m)^T$, the decision makers D_k $(k = 1, 2, \ldots, h)$ assign the HFLPRs $B_x^k = (b_{x,ij}^k)_{n\times n}$ to express the judgments of alternatives with respect to each criterion for $i = 1, 2, \ldots, n$, $x = 1, 2, \ldots, m$ and $k = 1, 2, \ldots, h$. A procedure for the Decision making process based on clustering and consensus can be concluded.

Algorithm 7.3 [7]. **A group Decision-Making procedure with HFLPRs.**

Input: The individual HFLPRs $B_x^k = (b_{x,ij}^k)_{n\times n}$ $(i = 1, 2, \ldots, n, x = 1, 2, \ldots, m,$ $k = 1, 2, \ldots, h)$, the free parameter γ, the number of clusters c, a constant value ξ, the maximum modified extent ρ, and the parameter function $f(d(\tau))$.

Output: The overall preference degrees of alternatives in each cluster.

Step 1. Aggregate all HFLPRs given by a decision maker into an overall one according to the weight vector $w = (w_1, w_2, \ldots, w_m)^T$, denoted as $B^k = (b_{ij}^k)_{n\times n}$ $(k = 1, 2, \ldots, h)$.

Step 2. Apply Algorithm 7.1 to classify decision makers into c clusters based on $B^k = (b_{ij}^k)_{n\times n}$ $(k = 1, 2, \ldots, h)$.

Step 3. Utilize Algorithm 7.2 to obtain the individual HFLPRs with acceptable consensus with respect to each criterion in each cluster.

Step 4. Aggregate all individual HFLPRs according to the criteria weights into an overall one in each cluster.

Step 5. Calculate and output the overall preference degrees of alternatives by the HFLWA in each cluster.

End.

Ren et al. [7] noted that classifying HFLPRs into clusters and obtaining Decision making results in each cluster may be helpful for enhancing the accuracy of the results. Furthermore, addressing the consensus issue for HFLPRs in each cluster may contribute to the reliability of the ranking even if decision makers have similar expertise.

7.4 Discussions

This section presents the convergence of the consensus process Algorithm 7.2 discussed by Ren et al. [7].

7.4.1 Convergence with Different Numbers of Decision Makers and Different Orders of HFLPRs

Ren et al. [7] designed the experiment as follows:

(1) Suppose that a group Decision making problem contains three objectives, then randomly generate three HFLPRs representing the judgments of the objectives given by three decision makers.

(2) Use Algorithm 7.2 to solve the Decision making problem in three different situations, i.e., the decision makers are insensitive, moderately sensitive, and sensitive to the distance between the individual HFLPRs and the overall HFLPR. Suppose that the parameter functions $f(d(\tau))$ are selected as: $f(d(\tau)) = 1 - (d(\tau))^2$, $f(d(\tau)) = 1 - d(\tau)$ and $f(d(\tau)) = 1 - (1 - (1 - d(\tau))^2)^{\frac{1}{2}}$, respectively, where $d(\tau)$ is the distance between an individual HFLPR and the overall HFLPR at the τth round.

(3) Assign the maximum modified extent $\rho = 0.1$, and get the ranking of objectives with the acceptable group consensus.

(4) Repeat (1) to (3) for 1000 times, record the average regulating times for achieving group consensus with decision makers' different sensitive attitudes. If a problem is processed with two rounds of HFLPR adjusting, then the regulating time is recorded as 2.

(5) Let the number of decision makers be four to six, and the number of objectives be four to seven. Repeat (1) to (4).

Some results can be got from the experiment [7]:

(1) When decision makers are insensitive or moderately sensitive to the distance between the individual HFLPRs and the overall HFLPR, the average regulating times for the 1000 Decision making problems remain 1 with different numbers of decision makers.

(2) When decision makers are insensitive or moderately sensitive to the distance between the individual HFLPRs and the overall HFLPR, the average regulating times for the 1000 Decision making problems remain 1 with different numbers of objectives.

(3) When decision makers are sensitive to the distance between the individual HFLPRs and the overall HFLPRs, the average regulating times for the 1000 Decision making problems are generally greater than 1.

(4) When decision makers are sensitive to the distance between the individual HFLPRs and the overall HFLPRs, for a certain number of decision makers, the average regulating times for the 1000 Decision making problems decrease with the increase of HFLPRs' orders. Table 7.1 presents the details (Table 7.2).

For immediacy, the results in Table 7.1 can be shown in Fig. 7.1.

Table 7.1 Variation of the average regulating times when decision makers hold sensitive attitude [7]

Average regulating times		The order of HFLPRs				
		3 * 3	4 * 4	5 * 5	6 * 6	7 * 7
The number of decision makers	3	1.705	1.663	1.596	1.563	1.553
	4	1.697	1.651	1.608	1.582	1.516
	5	1.710	1.649	1.619	1.568	1.555
	6	1.695	1.687	1.635	1.599	1.534

Table 7.2 The HFLPRs on the flood discharge of hydropower stations [7]

HFLPRs given by decision maker D_1	$B_1^1 = \begin{bmatrix} \{s_4\} & \{s_5, s_6\} & \{s_2, s_3, s_4\} \\ & \{s_4\} & \{s_2, s_3\} \\ & & \{s_4\} \end{bmatrix}$	$B_1^2 = \begin{bmatrix} \{s_4\} & \{s_4, s_5\} & \{s_3, s_4\} \\ & \{s_4\} & \{s_3, s_4, s_5\} \\ & & \{s_4\} \end{bmatrix}$
	$B_1^3 = \begin{bmatrix} \{s_4\} & \{s_3\} & \{s_3\} \\ & \{s_4\} & \{s_4, s_5, s_6\} \\ & & \{s_4\} \end{bmatrix}$	
HFLPRs given by decision maker D_2	$B_2^1 = \begin{bmatrix} \{s_4\} & \{s_4, s_5\} & \{s_3, s_4\} \\ & \{s_4\} & \{s_3, s_4\} \\ & & \{s_4\} \end{bmatrix}$	$B_2^2 = \begin{bmatrix} \{s_4\} & \{s_5, s_6\} & \{s_4, s_5\} \\ & \{s_4\} & \{s_4, s_5\} \\ & & \{s_4\} \end{bmatrix}$
	$B_2^3 = \begin{bmatrix} \{s_4\} & \{s_1, s_2\} & \{s_2, s_3\} \\ & \{s_4\} & \{s_3, s_4\} \\ & & \{s_4\} \end{bmatrix}$	
HFLPRs given by decision maker D_3	$B_3^1 = \begin{bmatrix} \{s_4\} & \{s_4, s_5, s_6\} & \{s_4, s_5\} \\ & \{s_4\} & \{s_2, s_3\} \\ & & \{s_4\} \end{bmatrix}$	$B_3^2 = \begin{bmatrix} \{s_4\} & \{s_5, s_6, s_7\} & \{s_4, s_5\} \\ & \{s_4\} & \{s_3\} \\ & & \{s_4\} \end{bmatrix}$
	$B_3^3 = \begin{bmatrix} \{s_4\} & \{s_2, s_3\} & \{s_3\} \\ & \{s_4\} & \{s_3\} \\ & & \{s_4\} \end{bmatrix}$	
HFLPRs given by decision maker D_4	$B_4^1 = \begin{bmatrix} \{s_4\} & \{s_5, s_6, s_7\} & \{s_4\} \\ & \{s_4\} & \{s_3, s_4\} \\ & & \{s_4\} \end{bmatrix}$	$B_4^2 = \begin{bmatrix} \{s_4\} & \{s_5, s_6\} & \{s_4, s_5\} \\ & \{s_4\} & \{s_4\} \\ & & \{s_4\} \end{bmatrix}$
	$B_4^3 = \begin{bmatrix} \{s_4\} & \{s_2, s_3\} & \{s_3\} \\ & \{s_4\} & \{s_4, s_5\} \\ & & \{s_4\} \end{bmatrix}$	
HFLPRs given by decision maker D_5	$B_5^1 = \begin{bmatrix} \{s_4\} & \{s_4, s_5\} & \{s_4, s_5\} \\ & \{s_4\} & \{s_4, s_5\} \\ & & \{s_4\} \end{bmatrix}$	$B_5^2 = \begin{bmatrix} \{s_4\} & \{s_5\} & \{s_4, s_5\} \\ & \{s_4\} & \{s_4\} \\ & & \{s_4\} \end{bmatrix}$
	$B_5^3 = \begin{bmatrix} \{s_4\} & \{s_1, s_2\} & \{s_2, s_3, s_4\} \\ & \{s_4\} & \{s_2, s_3\} \\ & & \{s_4\} \end{bmatrix}$	
HFLPRs given by decision maker D_6	$B_6^1 = \begin{bmatrix} \{s_4\} & \{s_4, s_5\} & \{s_4, s_5\} \\ & \{s_4\} & \{s_3, s_4\} \\ & & \{s_4\} \end{bmatrix}$	$B_6^2 = \begin{bmatrix} \{s_4\} & \{s_4, s_5\} & \{s_3, s_4\} \\ & \{s_4\} & \{s_3, s_4\} \\ & & \{s_4\} \end{bmatrix}$
	$B_6^3 = \begin{bmatrix} \{s_4\} & \{s_3, s_4\} & \{s_3, s_4\} \\ & \{s_4\} & \{s_3, s_4\} \\ & & \{s_4\} \end{bmatrix}$	

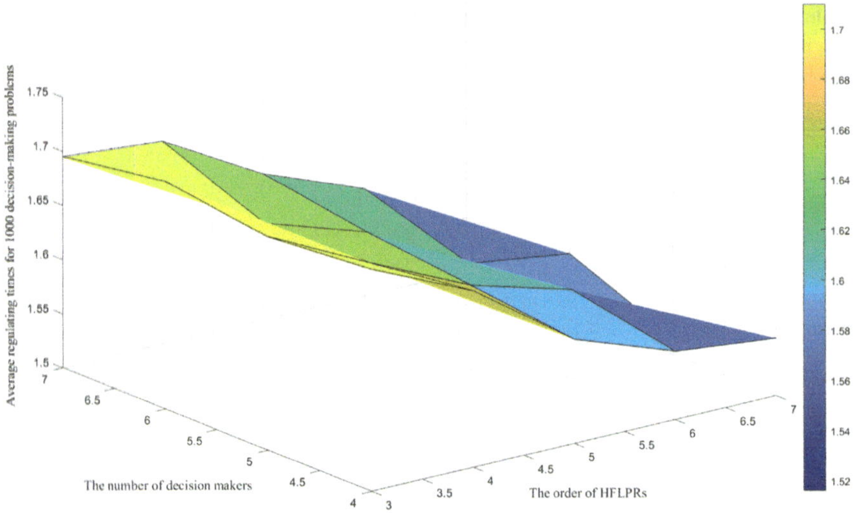

Fig. 7.1 Variation of average regulating times when the decision makers hold sensitive attitude [7]

Ren et al. [7] further concluded that: (1) if decision makers do not pay close attention to the distance between the individual HFLPRs and the overall HFLPR, the group consensus is likely to be achieved after one round of HFLPR adjusting, (2) decision makers with sensitive attitude probably need more feedback and adjustments, which is caused by their smaller compromises at each adjustment round; (3) the number of decision makers does not affect the adjustment procedure of Algorithm 7.2, which is a desirable property for large-scale Decision making problems.

7.4.2 Convergence with Different Maximum Modified Extents

This section randomly generates 1000 group Decision making problems with four objectives and three decision makers. Similarly, Ren et al. [7] handled the problems via Algorithm 7.2 with decision makers' different attitudes, i.e., insensitive, moderately sensitive, and sensitive. Ren et al. [7] determined the same functions of $f(d(\tau))$ in the above section, and set the maximum modified extent ρ as 0.02, 0.04, 0.06, and 0.08, respectively, then demonstrated the recorded average regulating times for achieving group consensus in Fig. 7.2.

By Fig. 7.2, with the increase of decision makers' sensibility, the average regulating times for achieving group consensus becomes increasingly sensitive to the change of maximum modified extent. Specifically, (1) when decision makers hold the moderately sensitive or sensitive attitude, the average regulating times decrease with the loose of maximum modified extent,(2) when decision makers hold an insensitive

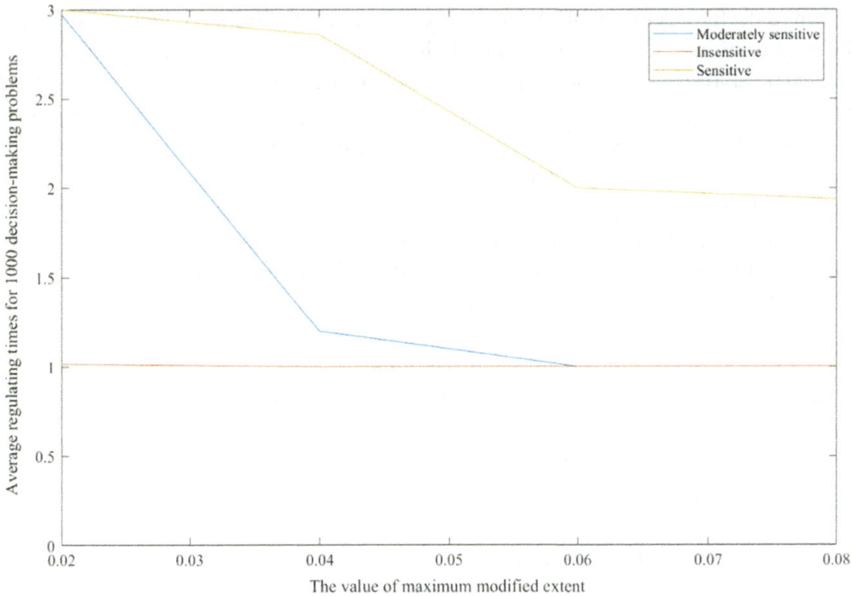

Fig. 7.2 Variation of average regulating times with different maximum modified extents [7]

attitude, the average regulating times are not influenced by the change of maximum modified extent [7].

Generally, the proposed consensus process is convergent under each Decision making situation considered in the above two parts [7].

7.5 A Case Study on Selecting an Optimal Flood Discharge Technique for a Hydropower Station

7.5.1 Description of the Case

As described in Chap. 6, the construction of the hydropower station has a lot of benefits, but it still affects the local environment. Flood discharge and energy dissipation, as a process for ensuring the safety of hydropower stations, usually brings adverse effects that cannot be ignored. Therefore, determining a suitable technology for flood discharge and energy dissipation can assure normal operation and minimize the environment's adverse impact. Commonly, the flood discharge technique is designed with considering (1) the depth of downstream water and the state of the riverbed to prevent soil erosion, (2) the water head level and atomization impacts to minimize humidity change in the drainage basin, among others [7].

Based on the necessity of selecting an effective flood discharge technique during the construction of a hydropower station, Ren et al. [7] simulated a Decision making process to address this kind of problem. Generally, there are three ways of flood discharge, i.e., underflow energy dissipation, trajectory bucket type energy dissipation, and roller bucket type energy dissipation. The advantages and disadvantages of the three ways can be briefly listed [7].

≫ *Underflow energy dissipation*:

Advantages: (1) stable flow state, (2) effective energy dissipation, (3) more adaptable for the geological conditions and the change of tailwater level;

Disadvantages: (1) high construction budget, (2) challenging to ensure the safety of the flooding pool itself for the high bottoming flow rate.

≫ *Trajectory bucket type energy dissipation*:

Advantages: (1) no required protection for the downstream channel, (2) low construction budget;

Disadvantages: (1) impacting the power station's output or navigation; (2) possible cavitation damage; (3) multiplying atomization effect.

≫ *Roller bucket type energy dissipation*:

Advantages: beneficial to drifting wood and discharging ice;

Disadvantages: unfavorable for the bank slope stability, power station operation and, downstream shipping conditions.

Ren et al. [7] supposed that six decision makers from a water resource development company, a design institute, and government organizations related to water conservancy projects are involved in selecting a technique for the flood discharge of hydropower station to be built. The criteria are determined as water head and bed configuration, impacts on erosion and atomization, and economic benefit.

Based on the background, Ren et al. [7] proposed to apply Algorithm 7.3 to solve the problem. Suppose that decision makers give their HFLPRs based on the LTS $S = \{s_\alpha | \alpha = 0, 1, \ldots, 2\tau\}$, where $\tau = 4$. If we denote the three flood discharge techniques as $\{A_i | i = 1, 2, 3\}$, the criteria as $\{C_x | x = 1, 2, 3\}$ and the six decision makers as $\{D_k | k = 1, 2, \ldots, 6\}$, then the HFLPRs with respect to each criterion given by the six decision makers (denoted as $B_k^x = (b_{k,ij}^x)_{3\times3}, x = 1, 2, 3, k = 1, 2, \ldots, 6$) can be presented:

7.5.2 *Decision-Making Process*

Suppose that criteria have the same weight, then by Steps 1–2 in Algorithm 7.3, if let $m = 2$, then the HFLPRs assigned by six decision makers can be classified as, i.e., Cluster 1 with HFLPRs from decision makers D_1, D_2, D_4 and Cluster 2 with HFLPRs from decision makers D_3, D_5, D_6 [7].

Furthermore, in each cluster, if decision makers determine the maximum modified extent as $\rho = 0.1$, and they are sensitive to the distance between the individual HFLPRs and the overall HFLPR, then by Steps 3–4 in Algorithm 7.3, the overall HFLPR in each cluster can be obtained [7]:

Table 7.3 The overall preference degrees of three techniques of flood discharge in clusters [7]

	A_1	A_2	A_3
Cluster 1	{3.63, 3.89, 4.15}	{3.51, 3.67, 3.82}	{3.88, 4.03, 4.07}
Cluster 2	{3.67, 4.26, 4.37}	{3.44, 3.74, 4.04}	{3.96, 4, 4.44}

$$B_{Cluster1} = \begin{bmatrix} \{s_4\} & \{s_{3.78}, s_{4.67}, s_{4.78}\} & \{s_{2.89}, s_{3.11}, s_{3.78}\} \\ & \{s_4\} & \{s_{2.78}, s_{3.34}, s_{4.12}\} \\ & & \{s_4\} \end{bmatrix}$$

and

$$B_{Cluster2} = \begin{bmatrix} \{s_4\} & \{s_{3.56}, s_{4.44}, s_{4.67}\} & \{s_{3.45}, s_{4.33}, s_{4.44}\} \\ & \{s_4\} & \{s_3, s_{3.67}, s_{3.67}\} \\ & & \{s_4\} \end{bmatrix}$$

.

The overall preference degrees of the three techniques of flood discharge in each cluster can be computed in Table 7.3.

Table 7.3 shows that the roller bucket type energy dissipation in both clusters is the optimal choice for the hydropower station to be built. Ren et al. [7] further noted that it is possible that different clusters lead to different results. Further analysis on coordinating the outcomes can be made in such a situation, such as by weighted calculation, finding a third party for obtaining a unified result, etc.

7.6 Summary

To make the group decision making with HFLPRs more feasible, this chapter has introduced a method for group decision making with HFLPRs based on the modified extent measurement, which contributes to [7]:

(1) Develop a clustering algorithm based on a kernel function to classify HFLPRs by mapping information into a higher dimension space with good adaptability.
(2) Establish a consensus process based on the modified extent of individual HFLPRs. It considers the different knowledge backgrounds and cognitions of decision makers, avoids their biased judgments for the unknown aspects, and ensures the effectiveness of HFLPRs during group decision making.
(3) Provide a framework for group decision making with HFLPRs, which allows decision makers to use their habitual expressions, includes a clustering process for accuracy, and considers a consensus model with incomplete knowledge and cognitions of decision makers.

(4) Design experiments to discuss the convergence of the proposed consensus process with decision makers' different sensitive attitudes to the distance between the individual HFLPRs and the overall HFLPR under different Decision making scenarios. The experiments provide decision support for decision makers to evaluate the consensus efficiency and determine the acceptable maximum modified extents in a specific Decision making environment.

References

1. Chen, X., Peng, L.J., Wu, Z.B., Pedrycz, W.: Controlling the worst consistency index for hesitant fuzzy linguistic preference relations in consensus optimization models. Comput. Ind. Eng. **143**, 136423 (2020)
2. Dong, Y.C., Chen, X., Herrera, F.: Minimizing adjusted simple terms in the consensus reaching process with hesitant linguistic assessments in group decision making. Inf. Sci. **297**, 95–117 (2015)
3. Graves, D., Pedrycz, W.: Kernel-based fuzzy clustering and fuzzy clustering: a comparative experimental study. Fuzzy Sets Syst. **161**(4), 522–543 (2010)
4. Liao, H.C., Xu, Z.S., Zeng, X.J.: Distance and similarity measures for hesitant fuzzy linguistic term sets and their application in multi-criteria decision making. Inf. Sci. **271**, 125–142 (2014)
5. Liu, H.B., Jiang, J.: Optimizing consistency and consensus improvement process for hesitant fuzzy linguistic preference relations and the application in group decision making. Inf. Fusion **56**, 114–127 (2020)
6. Liu, N.N., He, Y., Xu, Z.S.: A new approach to deal with consistency and consensus issues for hesitant fuzzy linguistic preference relations. Appl. Soft Comput. **76**, 400–415 (2019)
7. Ren, P.J., Xu, Z.S., Wang, X.X. Zeng, X.-J.: Group decision making with hesitant fuzzy linguistic preference relations based on modified extent measurement. Expert Syst. Appl. **171**, 114235 (2021)
8. Wu, P., Zhu, J., Zhou, L., Chen, H.: Local feedback mechanism based on consistency-derived for consensus building in group decision making with hesitant fuzzy linguistic preference relations. Comput. Ind. Eng. **137**, 106001 (2019)
9. Wu, Z.B., Xu, J.P.: Managing consistency and consensus in group decision making with hesitant fuzzy linguistic preference relations. Omega **65**, 28–40 (2016)
10. Zhang, Z.M., Chen, S.M.: A consistency and consensus-based method for group decision making with hesitant fuzzy linguistic preference relations. Inf. Sci. **501**, 317–336 (2019)
11. Zhang, Z.M., Chen, S.M.: Group decision making based on acceptable multiplicative consistency and consensus of hesitant fuzzy linguistic preference relations. Inf. Sci. **541**, 531–550 (2020)
12. Zhang, Z.M., Chen, S.M.: Group decision making with hesitant fuzzy linguistic preference relations. Inf. Sci. **514**, 354–368 (2020)
13. Zhu, B., Xu, Z.S.: Consistency measures for hesitant fuzzy linguistic preference relations. IEEE Trans. Fuzzy Syst. **22**, 34–45 (2014)

Chapter 8
A Consensus Model for Hesitant Fuzzy Linguistic Preference Relation Based on Consistency Driven

As the group decision making with HFLPRs has significant research value. Some literature has grown up around the theme of the consensus model with HFLPRs (As presented in Chap. 2). It is worth mentioning that the consistency property has been further considered in the group decision making process to improve the reliability of the results. However, most of the literature fails to address the consistency threshold for group decision making, which may impact the results' effectiveness. To overcome this meekness, Ren et al. [4] developed a group decision making method considering the consistency of individual HFLPRs and the group consensus level and discussed the consistency threshold based on the defined consistency index for decision support, which is what we will introduce.

8.1 Consistency Index for a Hesitant Fuzzy Linguistic Preference Relation

8.1.1 Consistency Index

Based on the consistency definition of a HFLPR proposed in [5], Ren et al. [4] rewrote the concept for more detail as follows:

A HFLPR $B = (b_{ij})_{n\times n}$ based on $S = \{s_\alpha | \alpha = 0, 1, \ldots, 2\tau\}$, where $b_{ij} = \left\{b_{ij}^s | s = 1, 2, \ldots, l_{b_{ij}}\right\}$ ($l_{b_{ij}}$ is the number of LTs in b_{ij}), is perfectly consistent if

$$\min_{s=1,2,\ldots,l_{b_{ij}}} \left| \frac{I(b_{ij}^s)}{2\tau} - \frac{w_i}{w_i + w_j} \right| = 0, i, j = 1, 2, \ldots, n \tag{8.1}$$

P. Ren and Z. Xu, *Decision-Making Analyses with Thermodynamic Parameters and Hesitant Fuzzy Linguistic Preference Relations*, Studies in Fuzziness and Soft Computing 409, https://doi.org/10.1007/978-3-030-73253-0_8

where $I(b_{ij}^s)$ is the ordered lower index of b_{ij}^s, and $w = (w_1, w_2, \ldots, w_n)^T$ is the priority vector derived from B satisfying $\sum_{i=1}^{n} w_i = 1, w_i \geq 0, i = 1, 2, \ldots, n$.

Then, the deviation existing between an element of a HFLPR and its corresponding consistent HFLPR can be given as follows:

Definition 8.1 [4] For a HFLPR $B = (b_{ij})_{n \times n}$, where $b_{ij} = \left\{ b_{ij}^s | s = 1, 2, \ldots, l_{b_{ij}} \right\}$ ($l_{b_{ij}}$ is the number of LTs in b_{ij}), and $w = (w_1, w_2, \ldots, w_n)^T$ is the derived priority vector, then the elementary consistency deviation between the preference on the alternatives x_i and x_j is

$$e_{ij} = \min_{s=1,2,\ldots,l_{b_{ij}}} \left| \frac{I(b_{ij}^s)}{2\tau} - \frac{w_i}{w_i + w_j} \right| \tag{8.2}$$

Motivated by the idea that measures the consistency through the distance between a LPR and the correspondingly consistent LPR [1], the consistency index for a HFLPR based on the global consistency deviation between a HFLPR and its correspondingly consistent HFLPR can be introduced as:

Definition 8.2 [4] For a HFLPR $B = (b_{ij})_{n \times n}$, where $b_{ij} = \left\{ b_{ij}^s | s = 1, 2, \ldots, l_{b_{ij}} \right\}$ ($l_{b_{ij}}$ is the number of LTs in b_{ij}), suppose that the priority vector $w = (w_1, w_2, \ldots, w_n)^T$ is derived from B, then a hesitant fuzzy linguistic consistency index (HCI) can be defined as:

$$HCI = \frac{2}{n(n-1)} \sum_{i,j=1, i<j} e_{ij}^2 \tag{8.3}$$

The smaller HCI indicates that the higher consistency of the HFLPR. Before giving further discussions on the consistency judgment of a HFLPR, a model for deriving a priority vector is presented.

For an acceptably consistent HFLPR $B = (b_{ij})_{n \times n}$, where $b_{ij} = \left\{ b_{ij}^s | s = 1, 2, \ldots, l_{b_{ij}} \right\}$ ($l_{b_{ij}}$ is the number of LTs in b_{ij}), a programming model aiming to minimize the consistency deviation can be easily constructed:

Model 8.1 [4]

$$\min_{s=1,2,\ldots,l_{b_{ij}}} \sum_{i=1}^{n} \sum_{j=1, i \neq j}^{n} \left| \frac{I(b_{ij}^s)}{2\tau} - \frac{w_i}{w_i + w_j} \right|$$

$$s.t. \sum_{i=1}^{n} w_i = 1, w_i \geq 0, i = 1, 2, \ldots, n$$

Proposition 8.1 [4] Model 8.1 is solvable.

Proof

Ren et al. [4] let $p_{ij} = \frac{I(b_{ij}^{1})}{2\tau} or \cdots or \frac{I(b_{ij}^{l_{b_{ij}}})}{2\tau}$, and rewrote the objective function of Model 8.1 as $\min \sum_{i=1}^{n} \sum_{j=1, i \neq j}^{n} \left|(1-p_{ij})w_i - p_{ij}w_j\right|$, then transformed Model 8.1 into the following equivalent form:

Model 8.2 [4].

$$\min \sum_{i=1}^{n} \sum_{j=1, i \neq j}^{n} (x_{ij}^{+} + x_{ij}^{-})$$

$$s.t. \begin{cases} (1-p_{ij})w_i - p_{ij}w_j - x_{ij}^{+} + x_{ij}^{-} = 0 \\ \sum_{i=1}^{n} w_i = 1, w_i \geq 0, i = 1, 2, \ldots, n \\ x_{ij}^{+}, \ x_{ij}^{-} \geq 0, i, j = 1, 2, \ldots, n, i \neq j \end{cases}$$

Model 8.2 is solvable as it is a linear programming model.

8.1.2 Acceptable Threshold

Since $HCI = 0$ is actually challenging to be achieved, this section discusses the acceptable consistency level of a HFLPR.

Motivated by the fact that people's certain consistency tendency [3] and the description of differences between a LPR and the correspondingly consistent LPR [1], Ren et al. [4] supposed that e_{ij} $(i, j = 1, 2, \ldots, n)$ are normally distributed at mean 0 and standard deviation σ, i.e., $e_{ij} \sim N(0, \sigma^2)$, and got that $(\frac{n(n-1)}{2\sigma} \cdot HCI)^2$ is a Chi-square distribution with degrees of freedom $\frac{n(n-1)}{2}$, i.e., $(\frac{n(n-1)}{2\sigma} \cdot HCI)^2 \sim \chi^2(\frac{n(n-1)}{2})$. Later on, a hypothesis testing was made as [4]:

Null hypothesis $H_0 : \sigma^2 \leq \sigma_0^2$,
Alternative hypothesis $H_1 : \sigma^2 > \sigma_0^2$,

where α is the significant level and λ_α is the critical value. Based on this hypothesis testing, Ren et al. [4] deduced that the value of acceptably consistency threshold could be solved by:

$$\overline{HCI} = \frac{2\sigma_0}{n(n-1)}\sqrt{\lambda_\alpha} \tag{8.4}$$

8.2 Group Consensus with Hesitant Fuzzy Linguistic Preference Relations

8.2.1 Consensus Index

Ren et al. [4] introduced a consensus index that records the difference between the individual HFLPRs and the overall HFLPR as follows:

Definition 8.3 [4]. Let $B^k = (b_{ij}^k)_{n\times n}$ $(k = 1, 2, \ldots, h)$ be h HFLPRs, and $B = (b_{ij})_{n\times n}$ be the corresponding overall HFLPR $B = (b_{ij})_{n\times n}$, then the individual deviation index of B^k is

$$IDI_k = \frac{2}{n(n-1)} \sum_{i,j=1,i<j}^{n} d(b_{ij}^k, b_{ij}) \tag{8.5}$$

and the group consensus index of the h HFLPRs is

$$HLCI = 1 - \frac{2}{hn(n-1)} \sum_{k=1}^{h} \sum_{i,j=1,i<j}^{n} d(b_{ij}^k, b_{ij}) \tag{8.6}$$

where $d(b_{ij}^k, b_{ij})$ is the distance measure between b_{ij}^k and b_{ij}, for $i, j = 1, 2, \ldots, n$ and $k = 1, 2, \ldots, h$.

Notably, suppose that $\overline{HLCI}$ is the consensus threshold, then if $HLCI \geq \overline{HLCI}$, the group consensus is acceptable; otherwise, it is unacceptable [4].

8.2.2 Consensus Reaching Process Based on Consistency Measurement

Here, we present the consensus process with consistency measurement as follows:

Algorithm 8.1 [4]. Consensus reaching process based on consistency measurement. *Input:* h HFLPR $B^k = (b_{ij}^k)_{n\times n}$ $(k = 1, 2, \ldots, h)$, the acceptably consistent threshold $\overline{HCI}$, the consensus threshold $\overline{HLCI}$.

Output: The overall HFLPR $B = (b_{ij})_{n\times n}$ with acceptable consensus.

Step 1. Obtain the priorities from $B^k = (b_{ij}^k)_{n\times n}$ by Model 8.2 for $k = 1, 2, \ldots, h$;

Step 2. Calculate the consistency index HCI^k of the HFLPRs by Eq. (8.3) for $k = 1, 2, \ldots, h$,

Step 3. For each HFLPRs, if $HCI^k \leq \overline{HCI}$, then go to Step 5; otherwise, go Step 3;

Step 4. Find i^k and j^k corresponding $\max_{i,j=1,\ldots,n}\left\{\min_{z=1,\ldots,l_{b_{ij}^k}}\left|\frac{I((b_{ij}^k)^z)}{2\tau}-\frac{w_i^k}{w_i^k+w_j^k}\right|\right\}$, denoted as i^{k^*} and j^{k^*}, and revise the maximum value in b_{ij}^k as $I^-\left(\left(\frac{I((b_{ij}^k)^z)}{2\tau}\right)^{\alpha}\left(\frac{w_{i*}^k}{w_{i*}^k+w_{j*}^k}\right)^{1-\alpha}\right)$ $(0 \le \alpha \le 1)$, turn to Step 2;

Step 5. Aggregate all acceptable HFLPRs $B^k = (b_{ij}^k)_{n\times n}$ for $k = 1, 2, \ldots, h$ into an overall HFLPR $B = (b_{ij})_{n\times n}$ by an aggregation operator, continue Step 6;

Step 6. Compute the group consensus index $HLCI$ of the HFLPRs $B^k = (b_{ij}^k)_{n\times n}$ for $k = 1, \ldots, h$ by Eq. (8.6). If $HLCI < \overline{HLCI}$, go to Step 7,otherwise, end the algorithm and output $B = (b_{ij})_{n\times n}$;

Step 7. Find the maximum value of IDI_k for $k = 1, 2, \ldots, h$, where IDI_k is the individual deviation index of the kth HFLPR. Let $\overline{B}^k = (\overline{b}_{ij}^k)_{n\times n}$ replace $B^k = (b_{ij}^k)_{n\times n}$, where $\overline{b}_{ij}^k = \gamma b_{ij}^k + (1-\gamma)b_{ij}$. Return Step 1.

8.3 A Decision-Making Procedure

For a decision making problem with n criteria and m alternatives, h decision makers compare the criteria and alternatives in pair by HFLPRs, then the decision making procedure based on the consensus model with consistency driven can be briefly shown [4]:

Step 1. Put the HFLPRs over criteria, the acceptably consistent threshold $\overline{HCI}$ and the consensus threshold $\overline{HLCI}$ into Algorithm 6.1 to obtain the overall HFLPR. Go to Step 2;

Step 2. Calculate the priorities of the obtained HFLPR by Model 8.2.

Step 3. Repeat Steps 1 and 2 for the HFLPRs on alternatives with respect to each criterion. Continue Step 4.

Step 4. Aggregate the final priorities of alternatives according to the priorities of the criteria and alternatives obtained in Steps 2 and 3, and rank alternatives.

8.4 Discussions

8.4.1 Acceptably Consistent Threshold

Since acceptably consistent threshold $\overline{HCI} = \frac{2\sigma_0}{n(n-1)}\sqrt{\lambda_\alpha}$, then the key to determine $\overline{HCI}$ is to obtain the value of σ_0 for each order of HFLPR. To address this issue, Ren et al. [4] gave the following experiments:

Table 8.1 The suggested values of $\overline{HCI}$ for different orders of HLFPRs [4]

The order of HFLPR	The confidence interval of S_ϑ with the confidence level of 0.05	The reasonable interval of $\overline{HCI}$
$n = 3$	$[0.1497, 0.1527]$	$[0.1395, 0.1422]$
$n = 4$	$[0.1546, 0.1562]$	$[0.0914, 0.0924]$
$n = 5$	$[0.1581, 0.1597]$	$[0.0677, 0.0683]$
$n = 6$	$[0.1547, 0.1559]$	$[0.0516, 0.0520]$
$n = 7$	$[0.1435, 0.1440]$	$[0.0391, 0.0392]$
$n = 8$	$[0.1489, 0.1493]$	$[0.0342, 0.0343]$
$n = 9$	$[0.1574, 0.1578]$	$[0.0308, 0.0309]$
$n = 10$	$[0.1674, 0.1677]$	$[0.0292, 0.0293]$
$n = 11$	$[0.1779, 0.1783]$	$[0.0277, 0.0278]$
$n = 12$	$[0.1906, 0.1910]$	$[0.0266, 0.0267]$

(1) For the order of HFLPRs $n = 3$, we randomly generate 1000 HFLPRs, and get the priorities of the 1000 HFLPRs by Model 8.2;
(2) For each generated HFLPR, we calculate the elementary consistency deviations e_{ij}^x by Eq. (8.2) for $i, j = 1, 2, \ldots, n$ and $x = 1, 2, \ldots, 1000$, and their average variances, denoted as S^x. Furtherly, we calculate the average value and the average variance of S^x for $x = 1, 2, \ldots, 1000$.
(3) Let n be 4, 5, 6, 7, 8, 9, 10, 11, 12, we repeat (1) and (2), get the average value and the average variance of S^x for each order of HFLPRs;
(4) Use parameter estimation to acquire the confidence interval of S^x based on (3) (use Z statistic), and correspondingly compute the interval of $\overline{HCI}$.

The results obtained from the experiments are shown in Table 8.1 and Fig. 8.1.

By Table 8.1 and Fig. 8.1, Ren et al. [4] concluded that (1) the high order of matrices relatively possess small values of the lower limit and upper limit,(2) the lower limit approaches to the upper limit when the order of matrices increase.

8.4.2 *Applicability of the Proposed Procedure*

To discuss the applicability of the proposed method, for any random 1000 problems with h decision makers, Ren et al. [4] handled them through the following scenarios:

Scenarios 1. Let $h = 3$, and let the order of HFLPRs n assigned by decision makers be 4, 5, 6, 7, respectively, then apply the proposed procedure in Sect. 8.3 to solve the random HFLPRs in every 1000 problems, record the average execution time that gets the decision making results for each order of HFLPRs.

Scenarios 2. Extend the number of decision makers h to 4, 5, 6, do the same experiments as Scenarios 1.

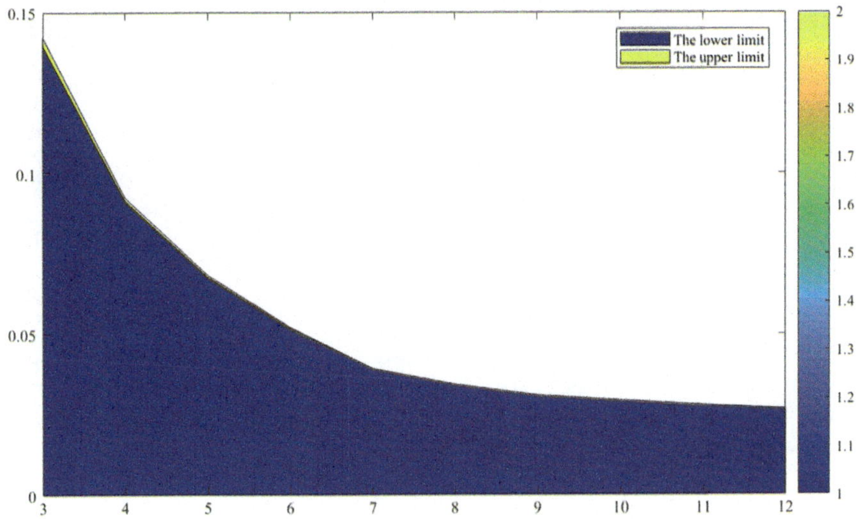

Fig. 8.1 The suggested values of $\overline{HCI}$ for different orders of HFLPRs [4]

Table 8.2 The average execution time of the proposed procedure with different numbers of decision makers and different orders of HFLPRs [4]

Average execution time	The order of HFLPRs			
	$n = 4$	$n = 5$	$n = 6$	$n = 7$
$h = 3$	0.8000	0	0	0
$h = 4$	2.3333	0	0	0
$h = 5$	3.6667	6	0.6667	4
$h = 6$	13.3333	13.3333	13.5000	10

The experiment results can be listed in Table 8.2.

The results in Table 6.2 can be visually shown in Figs. 8.2 and 8.3.

Ren et al. [4] indicated that the average execution time increases with the increase of decision makers due to that reconciling opinions among more people is more demanding than with a small number of people when they hold different judgments. This result demonstrates that the proposed procedure is more suitable to be applied to small-scale decision making problems.

Ren et al. [4] explained that the undefined variation of the average execution time in Fig. 8.3 might be caused by (1) the increase of objectives leading to more illogical individual judgments,(2) group analyses might lead that decision makers gradually figure out the comparative criteria with multiple comparisons when objectives increase.

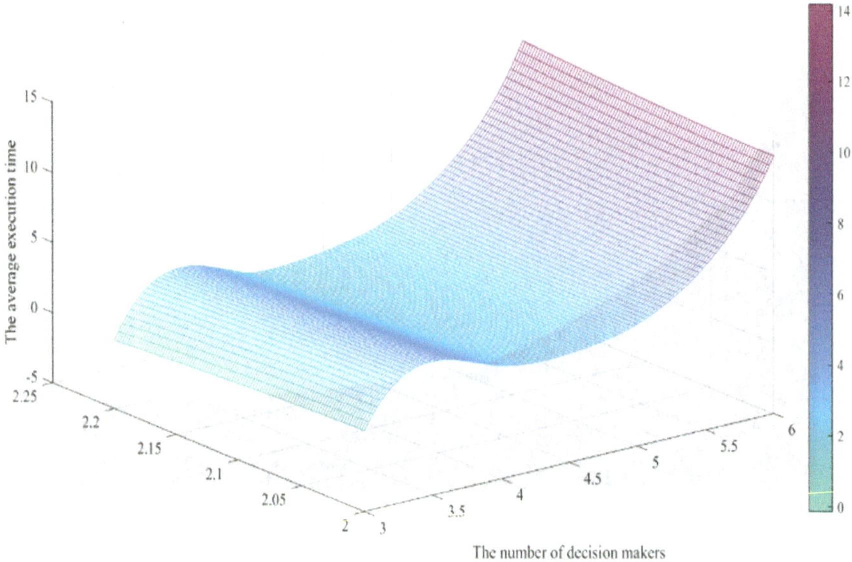

(1) The order of HFLPRs $n = 4$

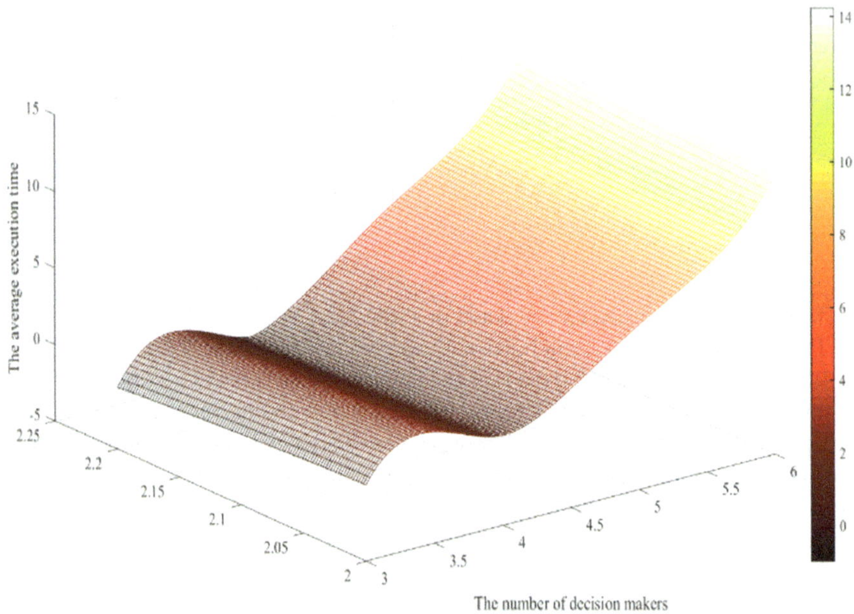

(2) The order of HFLPRs $n = 5$

Fig. 8.2 Variation of the average execution time with different orders of HFLPRs [4]

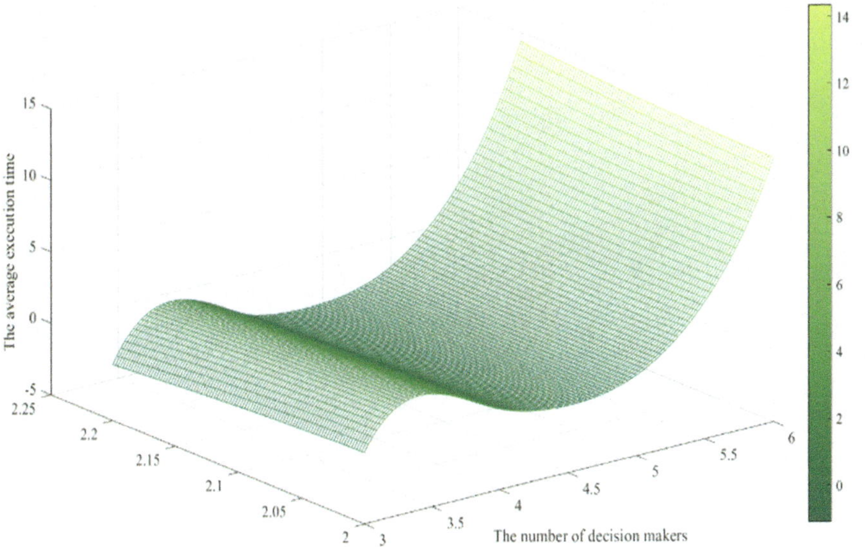

(3) The order of HFLPRs $n = 6$

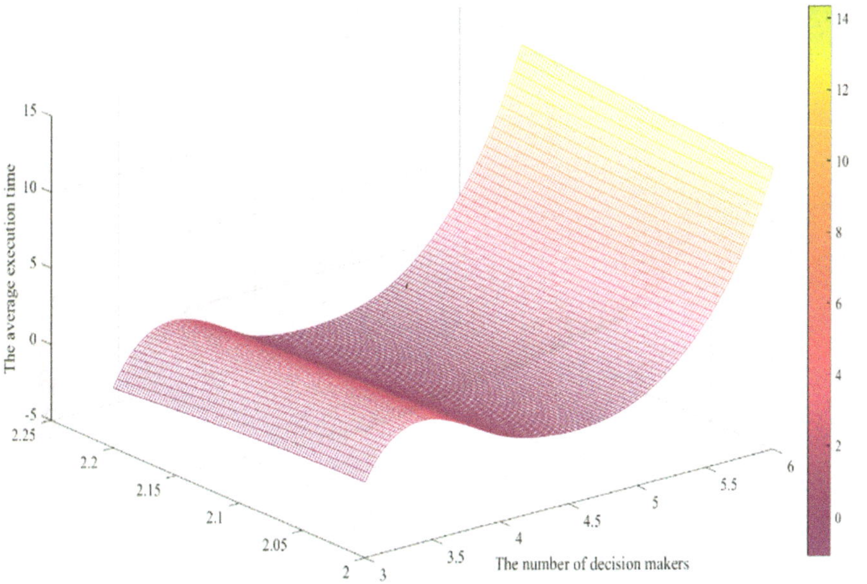

(4) The order of HFLPRs $n = 7$

Fig. 8.2 (continued)

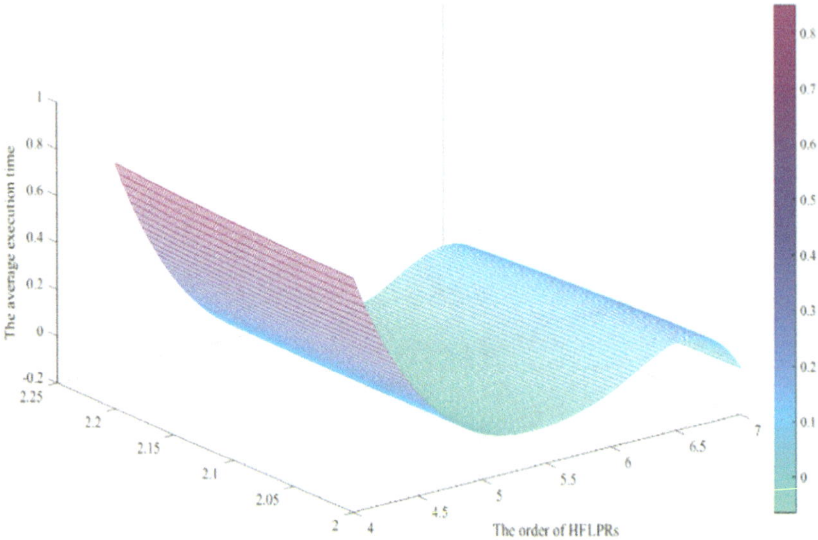

(1) The number of decision-makers $h = 3$

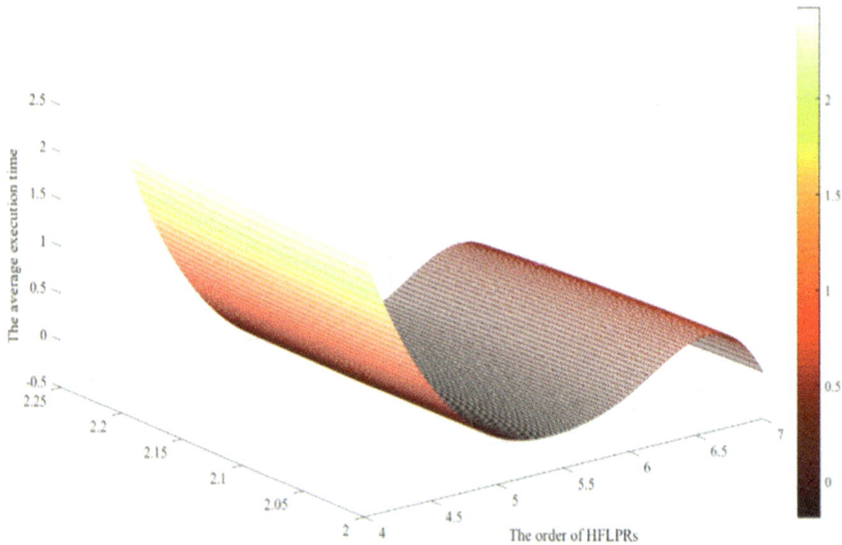

(2) The number of decision-makers $h = 4$

Fig. 8.3 Variation of the average execution time with different numbers of decision makers [4]

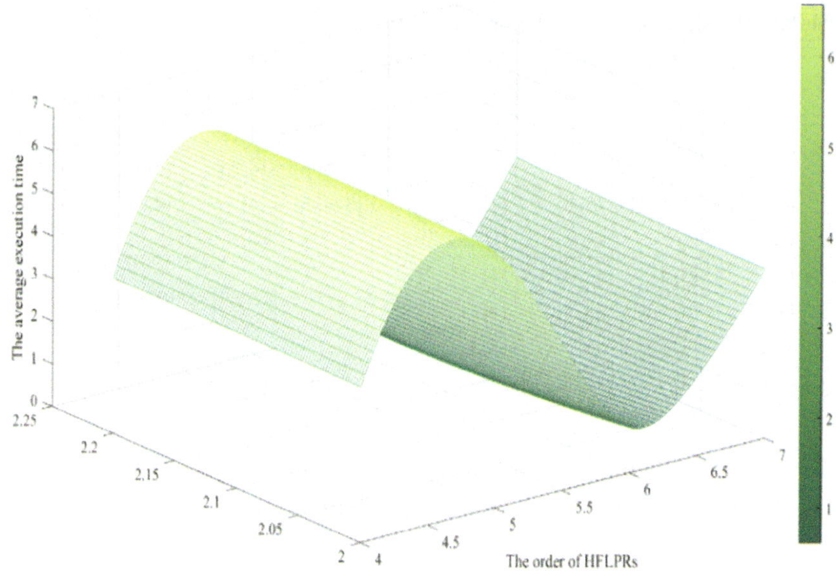

(3) The number of decision-makers $h = 5$

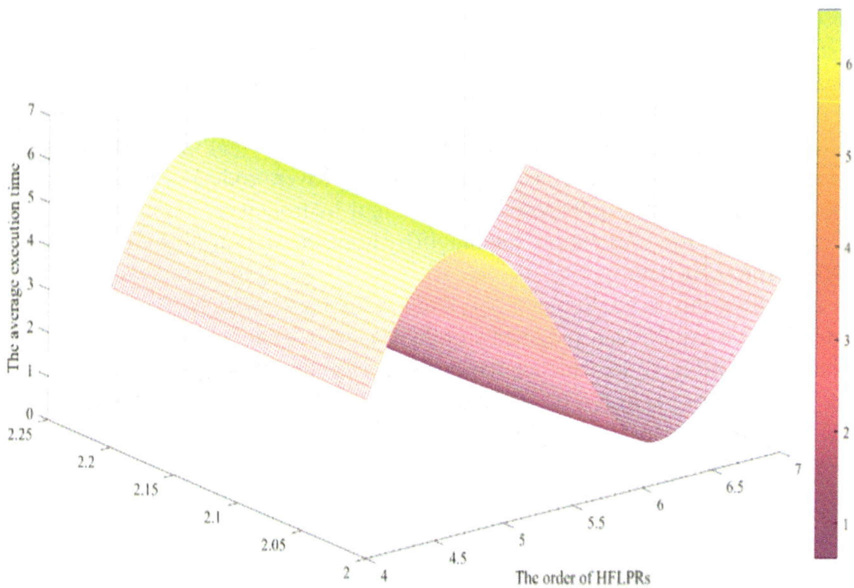

(4) The number of decision-makers $h = 6$

Fig. 8.3 (continued)

Table 8.3 Decision makers' HFLPRs on criteria [4]

k_1	C_1	C_2	C_3	C_4	k_2	C_1	C_2	C_3	C_4
C_1	{4}	{2, 3}	{3, 4}	{4, 5, 6}	C_1	{4}	{2, 3, 4}	{2, 3, 4}	{3, 4, 5}
C_2	{5, 6}	{4}	{4}	{2, 3, 4}	C_2	{4, 5, 6}	{4}	{3, 4}	{4}
C_3	{4, 5}	{4}	{4}	{2, 3}	C_3	{4, 5, 6}	{4, 5}	{4}	{3, 4}
C_4	{2, 3, 4}	{4, 5, 6}	{5, 6}	{4}	C_4	{3, 4, 5}	{4}	{4, 5}	{4}
k_3	C_1	C_2	C_3	C_4					
C_1	{4}	{3, 4}	{3, 4}	{4}					
C_2	{4, 5}	{4}	{3, 4}	{4}					
C_3	{4, 5}	{4, 5}	{4}	{3, 4, 5}					
C_4	{4}	{4}	{3, 4, 5}	{4}					

8.5 A Case Study on Assessing the Erosion Impacts of Hydropower Stations on Environment

8.5.1 *Description of the Case*

Based on the descriptions in Chap. 7 that flood discharge and energy dissipation of a hydropower station cause adverse effects on the environment: shaking, atomization, erosion, and aeration, Ren et al. [4] further discussed the erosion effect of a hydropower station and determined the assessing criteria as (1) downstream bed formation lithology,(2) the development of discontinuity; (3) the quality of rock mass; (4) flow [2]. Later on, considering that Jinsha River produces lots of electricity every year, Ren et al. [4] proposed to assess four hydropower stations in this area: Lianghekou hydropower station, Jinpingyiji hydropower station, Guandi hydropower station, Tongzilin hydropower station.

To address this issue, Ren et al. [4] supposed three experts to be invited to give judgments on criteria and judgments on hydropower stations with respect to each criterion according to the characteristics and performances of hydropower stations. Denoting the criteria as C_1, C_2, C_3 and C_4, the hydropower stations A_1, A_2, A_3 and A_4, and the experts as D_1, D_2 and D_3, the HFLPRs assigned by experts can be listed as follows Tables 8.3, 8.4, 8.5, 8.6 and 8.7:

8.5.2 *Decision-Making Process*

By applying the proposed decision making procedure to handle the experts' HFLPRs, the following results can be obtained [4]:

(1) The priority vector of criteria:

Table 8.4 Decision makers' HFLPRs on water conservancy projects with respect to C_1 [4]

k_1	C_1	C_2	C_3	C_4	k_2	C_1	C_2	C_3	C_4
C_1	{4}	{2, 3}	{3, 4}	{4, 5, 6}	C_1	{4}	{2, 3, 4}	{2, 3, 4}	{3, 4, 5}
C_2	{5, 6}	{4}	{4}	{2, 3, 4}	C_2	{4, 5, 6}	{4}	{3, 4}	{4}
C_3	{4, 5}	{4}	{4}	{2, 3}	C_3	{4, 5, 6}	{4, 5}	{4}	{3, 4}
C_4	{2, 3, 4}	{4, 5, 6}	{5, 6}	{4}	C_4	{3, 4, 5}	{4}	{4, 5}	{4}
k_3	C_1	C_2	C_3	C_4					
C_1	{4}	{3, 4}	{3, 4}	{4}					
C_2	{4, 5}	{4}	{3, 4}	{4}					
C_3	{4, 5}	{4, 5}	{4}	{3, 4, 5}					
C_4	{4}	{4}	{3, 4, 5}	{4}					

Table 8.5 Decision makers' HFLPRs on water conservancy projects with respect to C_2 [4]

k_1	C_1	C_2	C_3	C_4	k_2	C_1	C_2	C_3	C_4
C_1	{4}	{2, 3}	{3, 4}	{4, 5, 6}	C_1	{4}	{2, 3, 4}	{2, 3, 4}	{3, 4, 5}
C_2	{5, 6}	{4}	{4}	{2, 3, 4}	C_2	{4, 5, 6}	{4}	{3, 4}	{4}
C_3	{4, 5}	{4}	{4}	{2, 3}	C_3	{4, 5, 6}	{4, 5}	{4}	{3, 4}
C_4	{2, 3, 4}	{4, 5, 6}	{5, 6}	{4}	C_4	{3, 4, 5}	{4}	{4, 5}	{4}
k_3	C_1	C_2	C_3	C_4					
C_1	{4}	{3, 4}	{3, 4}	{4}					
C_2	{4, 5}	{4}	{3, 4}	{4}					
C_3	{4, 5}	{4, 5}	{4}	{3, 4, 5}					
C_4	{4}	{4}	{3, 4, 5}	{4}					

Table 8.6 Decision makers' HFLPRs on water conservancy projects with respect to C_3 [4]

k_1	C_1	C_2	C_3	C_4	k_2	C_1	C_2	C_3	C_4
C_1	{4}	{2, 3}	{3, 4}	{4, 5, 6}	C_1	{4}	{2, 3, 4}	{2, 3, 4}	{3, 4, 5}
C_2	{5, 6}	{4}	{4}	{2, 3, 4}	C_2	{4, 5, 6}	{4}	{3, 4}	{4}
C_3	{4, 5}	{4}	{4}	{2, 3}	C_3	{4, 5, 6}	{4, 5}	{4}	{3, 4}
C_4	{2, 3, 4}	{4, 5, 6}	{5, 6}	{4}	C_4	{3, 4, 5}	{4}	{4, 5}	{4}
k_3	C_1	C_2	C_3	C_4					
C_1	{4}	{3, 4}	{3, 4}	{4}					
C_2	{4, 5}	{4}	{3, 4}	{4}					
C_3	{4, 5}	{4, 5}	{4}	{3, 4, 5}					
C_4	{4}	{4}	{3, 4, 5}	{4}					

Table 8.7 Decision makers' HFLPRs on water conservancy projects with respect to C_4 [4]

k_1	C_1	C_2	C_3	C_4	k_2	C_1	C_2	C_3	C_4
C_1	{4}	{2, 3}	{3, 4}	{4, 5, 6}	C_1	{4}	{2, 3, 4}	{2, 3, 4}	{3, 4, 5}
C_2	{5, 6}	{4}	{4}	{2, 3, 4}	C_2	{4, 5, 6}	{4}	{3, 4}	{4}
C_3	{4, 5}	{4}	{4}	{2, 3}	C_3	{4, 5, 6}	{4, 5}	{4}	{3, 4}
C_4	{2, 3, 4}	{4, 5, 6}	{5, 6}	{4}	C_4	{3, 4, 5}	{4}	{4, 5}	{4}
k_3	C_1	C_2	C_3	C_4					
C_1	{4}	{3, 4}	{3, 4}	{4}					
C_2	{4, 5}	{4}	{3, 4}	{4}					
C_3	{4, 5}	{4, 5}	{4}	{3, 4, 5}					
C_4	{4}	{4}	{3, 4, 5}	{4}					

$$w_C = (0.2851, 0.2340, 0.1497, 0.3312)^T \text{ with } HLCI_C = 0.9547;$$

(2) The priority vector of the hydropower stations with respect to each criterion:

with respect to C_1, $w_1 = (0.3385, 0.2357, 0.0923, 0.3335)^T$ with $HLCI_1 = 0.9280$;
with respect to C_2, $w_2 = (0.2676, 0.1714, 0.3117, 0.2492)^T$ with $HLCI_2 = 0.9425$;
with respect to C_3, $w_3 = (0.3019, 0.1470, 0.2728, 0.2783)^T$ with $HLCI_3 = 0.9547$;
with respect to C_4, $w_4 = (0.2173, 0.2642, 0.3108, 0.2076)^T$ with $HLCI_4 = 0.9026$.

Based on the weighted averaging operator, the final scores for the four hydropower stations can be computed as 0.2759, 0.2456, 0.2535, and 0.2391, and the ranking of them is [4]

$$Lianghekou \succ Guandi \succ Jinpingyiji \succ Tongzilin.$$

The result shows that the Lianghekou hydropower station is friendliest toward the erosion impact on the environment.

8.5.3 Comparisons

For comparison, Ren et al. [4] applied the method introduced by [6] to solve the same case. The results are:

(1) The priority vector of criteria:

$$w_C = (0.2433, 0.2767, 0.2433, 0.2367)^T;$$

(2) The priority vector of the hydropower stations with respect to C_1, C_2, C_3, C_4:

$$w_1 = (0.2500, 0.2400, 0.2467, 0.2633)^T;$$
$$w_2 = (0.2325, 0.2375, 0.2650, 0.2650)^T;$$
$$w_3 = (0.2650, 0.2317, 0.2450, 0.2583)^T;$$
$$w_4 = (0.2400, 0.3033, 0.2300, 0.2267)^T.$$

The ranking of the hydropower stations is

$$Tongzilin \succ Jinpingyiji \succ Guandi \succ Lianghekou,$$

which is different from the results obtained by the proposed decision making procedure. Based on that the method proposed by [6] and the method proposed by [7] (with consistency property) are non-differential, Ren et al. [4] deduced that the differences caused in the case might be the consensus consideration.

8.6 Summary

This chapter has presented a procedure for HFLPRs with individual consistency and group consensus, which has the below contributions [4]:

(1) It allows decision makers to express judgments based on their natural expression habits.
(2) Based on addressing the consistency of individual preference relation and the consensus of group preference relation, it guarantees reliable and effective results.
(3) It discusses the consistency threshold for each order of HFLPRs to support the procedure through the meaning of errors rather than the typical value 0.9, which improves its availability and applicability.
(4) It repairs the HFLPRs during consistency improvement by revising the most unharmonious element, which significantly retains decision makers' original judgments.

Finally, Ren et al. [4] put forward to make extensions of the proposed procedure into a more general environment, such as the problem involving both subjective and objective criteria,the situation considering incomplete information, etc.

References

1. Dong, Y.C., Xu, Y.F., Li, H.Y.: On consistency measures of linguistic preference relations. Eur. J. Oper. Res. **189**(2), 430–444 (2008)

2. Huang, Z.W., Lin, S.X., Huang, H.J., Xu, G.M.: Analyzing the factors and evolutions of the downstream scour for Danjiangkou dam. Yangzi River **38**(9), 34–37 (2007)
3. Jong, P.: A statistical approach to Saaty's scaling method for priorities. J. Math. Psychol. **28**(4), 467–478 (1984)
4. Ren, P. J., Wang, X. X., Zeng, X.-J. & Xu, Z. S.: Research on consistency and consensus reaching models for hesitant fuzzy linguistic preference relation. *Technical Report* (2020)
5. Ren, P.J., Zhu, B., Xu, Z.S.: Assessment of the impact of hydropower stations on the environment with a hesitant fuzzy linguistic hyperplane-consistency programming method. IEEE Trans. Fuzzy Syst. **26**(5), 2981–2992 (2018)
6. Wu, Z.B., Xu, J.P.: Managing consistency and consensus in group decision making with hesitant fuzzy linguistic preference relations. Omega **65**, 28–40 (2016)
7. Zhang, Z.M., Wu, C.: On the use of multiplicative consistency in hesitant fuzzy linguistic preference relations. Knowl.-Based Syst. **72**, 13–27 (2014)

Chapter 9
Conclusions and Outlooks

9.1 Conclusions

To make full use of decision making information and apply efficient information expression way to improve the accuracy of decision making, this paper has shown the decision making methods with thermodynamic parameters and HFLPRs. The main work of the book can be summarized as follows:

(1) In Chap. 2, we systematically review the existing literature about the decision making models with thermodynamics and the decision making methods with HFLPRs based on the database of Web of Science, Google Scholar, etc.

(2) Introduce the thermodynamic parameters under different uncertain environment and decision making methods based on thermodynamic parameters under different types of information. Specifically,

    (1) Present the intuitionistic decision energy, intuitionistic decision exergy, intuitionistic decision entropy, and other concepts to depict the intuitionistic information from the perspectives of numerical value and distribution, and then build a method for intuitionistic decision making with IFSs in Chap. 3. Furthermore, comparative analyses among the given method, the IFWAO, and the IF-TOPSIS have been made through experiments. The comparisons show that the distribution of decision making information is a factor to impact the results, which verifies the scientificity and effectiveness of the given method.

    (2) Based on constructing the hesitant fuzzy matrix with prospect theory and introducing the thermodynamic parameters in a hesitant fuzzy environment, a thermodynamic decision method based on prospect theory in a hesitant fuzzy environment has been provided in Chap. 4. The advantages of the methods are (I) allowing a general information expression for decision makers; (II) portraying different attitudes of decision makers towards the pros and cons of alternatives; (3) applying the features

P. Ren and Z. Xu, *Decision-Making Analyses with Thermodynamic Parameters and Hesitant Fuzzy Linguistic Preference Relations*, Studies in Fuzziness and Soft Computing 409, https://doi.org/10.1007/978-3-030-73253-0_9

of thermodynamic parameters and prospect theory to get accurate and reasonable decision making results. Comparisons have demonstrated the validity of the method among other methods and time complexity studies, which show that the given method can be used to obtain a deliberate ranking in a relatively simple way.

(3) Considering a problem may contain objective and subjective criteria and people prefer to use the natural linguistic expression for evaluating, Chap. 5 has given a thermodynamic decision making method for heterogeneous forms including real numbers, utility values, linguistic information, and hesitant fuzzy linguistic information. Meanwhile, since different fuzzy degrees of linguistic descriptions indicate the different confidences of decision makers on judgments, the correction process of criteria weights has been displayed to reduce the evaluations' criteria weights with high uncertainty. Moreover, decision makers' sensitivity to the accuracy of decision information has been discussed during the weight correction process to provide decision support for the given method.

(3) Provide consistency index and group consensus index for HFLPRs, and introduce the decision models and consensus mechanisms for decision making with HFLPRs. Specifically,

(1) With hyperplane theory for representing the weight vector plane, a mathematical programming model has been presented in Chap. 6 to derive the weight vector based on the decision maker's satisfaction measurement. The discussions on the unknown parameter in the given model have been made, and the suggestions of possible values for the parameter have been shown. In addition, the feasibility and validity of the given model have been manifested by comparative analyses with a method addressing consistent HFLPRs. The analyses testify that the optimal selections obtained by the given model are non-differential with the method addressing consistent HFLPRs.

(2) Since decision makers involving a problem may have different expertise and compromising according to the group opinions sometimes may exceed their own knowledge, a consensus model for group decision making with HFLPRs has been introduced in Chap. 7. It involves a clustering algorithm based on kernel function for clustering decision makers with different knowledge backgrounds and a consensus index based on the individual modification measurement with the limitation of thinking and cognition of decision makers. Compared with the other consensus decision making model, this given model corresponds well to decision makers' psychological behaviors during the decision making process. Furthermore, consensus efficiency and the acceptable maximum modified extents have been discussed in different decision making scenarios to make the given method applicable.

(3) To provide an effective way for group decision making with HFLPRs, based on the consistency definition and consensus definition, Chap. 8 has demonstrated a group consensus model for HFLPRs with consistency driven. Notably, the consistency thresholds for the given consistency index have been discussed in different decision making scenarios, which shows that the higher order of HFLPRs corresponds to the smaller acceptable consistency threshold. Besides, the applicability of the given model has been manifested through the complexity analysis, which indicates that it has an excellent ability to deal with small-scale decision making problems.

9.2 Outlooks

Based on the analysis of the current work and the introduction of the book, there still exist some issues to be concerned about in the future.

9.2.1 Future Work for Decision-Making Methods Based on Thermodynamic Parameters

(1) Construct the weight-determination mechanism. In the current research, criteria weights are determined by decision makers according to their experiences, which may lead to deviation from the objective situation. Therefore, it is worthy of introducing machine learning into uncertain decision making problems to obtain criteria weights based on historical data.

(2) Study the thermodynamic decision models based on bounded rationality. Since people are inevitably bounded rationality in economic management activity [4], it is necessary to capture decision makers' behavior patterns in decision making. More detailly, the decision making methods combining the thermodynamic parameters and decision makers' psychological characteristics, such as prospect theory [5], Dunning-Kruger effect [2], etc., should be systematically studied.

(3) Investigate the dynamic decision making method based on thermodynamic parameters. The problems we meet in practice are probably multi-stage, such as supply chain decision making, venture capital decision making, etc. Therefore, based on the existing decision making theory and the extraction of dynamic decision making elements and characteristics, the multi-stage decision making methods based on thermodynamic parameters are worth establishing.

(4) Develop the thermodynamic decision models with rough sets. Since decision making with rough set is an interesting area [1, 3, 6, 7], it is necessary to present methods for rough sets with thermodynamic parameters to improve the applicability of thermodynamic decision making.

9.2.2 Future Work for Decision-Making Methods Based on Hesitant Fuzzy Linguistic Preference Relations

(1) Develop novel priority derivation models. Most priority derivation models were proposed based on the consistency of HFLPRs. To make the theory fruitful, relying on different mathematical approaches to establish the priority derivation model and developing the decision support system and application software is feasible.

(2) Address the ranking theory based on different consistency definition. Currently, the discussion on the acceptable consistency level based on the existing consistency index lacks strict mathematical proof, which may lead to the irrationality of decision making results or increase the time complexity of decision making. Therefore, in view of the different consistency definition of HFLPRs, the relevant thresholds should be discussed.

(3) Build the decision theory of HFLPRs by considering decision makers' limited thinking. Based on decision makers' bounded rationality, it is meaningful to introduce prospect theory into the decision making with HFLPRs and provide the psychological perception index. Moreover, the consistency index, weight derivation mechanism, and consensus reaching model for HFLPRs with the bounded rationality can be achieved.

It is also significant to further broaden the research direction and put forward innovative theories in the field of uncertain decision making based on mathematics, computer, management, economics, and other disciplines in the future.

References

1. Fedrizzi, M., Kacprzyk, J., Nurmi, H.: How different are social choice functions: a rough sets approach. Qual. Quant. **30**, 87–99 (1996)
2. Kruger, J., Dunning, D.: Unskilled and unaware of it: how difficulties in recognizing one's own incompetence lead to inflated self-assessments. J. Pers. Soc. Psychol. **77**(6), 1121–1134 (1999)
3. Nurmi, H., Kacprzyk, J., Fedrizzi, M.: How different are social choice functions: a rough sets approach. Eur. J. Oper. Res. **95**(2), 264–277 (1996)
4. Simon, H.A.: Administrative Behavior: A Study of Decision-Making Processes in Administrative Organization. Macmillan Company, New York (1947)
5. Tversky, A., Kahneman, D.: Advances in prospect theory: cumulative representation of uncertainty. J. Risk Uncertain. **5**, 297–323 (1992)
6. Yao, Y., Greco, S., Sowiński, R.: Probabilistic Rough Sets. Springer Handbook of Computational Intelligence. Springer Berlin Heidelberg (2015)
7. Zhang, H., Shu, L., Xiong, L.: On novel hesitant fuzzy rough sets. Soft. Comput. **23**, 11357–11371 (2019)

The manufacturer's authorised representative in the EU is Springer Nature Customer Service Centre GmbH, Europaplatz 3, 69115 Heidelberg, Germany. If you have any concerns regarding our products, please contact ProductSafety@springernature.com

Printed and bound by CPI Group (UK) Ltd, Croydon, CR0 4YY
28/09/2026
02224086-0016